AUTHENTICITY IN PRESERVATION OF HISTORICAL WOODEN
ARCHITECTURE - PROBLEMS AND CHALLENGES

Authenticity in preservation of Historical Wooden Architecture - Problems and Challenges

Case Studies from the American South

Tomasz Tomaszek

Department of Monument Conservation, Faculty of Civil and Environmental Engineering and Architecture, Rzeszow University of Technology; Rzeszów, Poland

CRC Press
Taylor & Francis Group
Boca Raton London New York

CRC Press is an imprint of the
Taylor & Francis Group, an **informa** business

A BALKEMA BOOK

CRC Press/Balkema is an imprint of the Taylor & Francis Group, an informa business

© 2020 Taylor & Francis Group, London, UK

Typeset by Integra Software Services Pvt. Ltd., Pondicherry, India

Published by: CRC Press/Balkema
Schipholweg 107C, 2316XC Leiden, The Netherlands

First issued in paperback 2023

ISBN: 978-1-03-257101-0 (pbk)
ISBN: 978-0-367-46163-8 (hbk)
ISBN: 978-1-003-02732-4 (ebk)

DOI: https://doi.org/10.1201/9781003027324

Publisher's Note
The publisher has gone to great lengths to ensure the quality of this reprint but points out that some imperfections in the original copies may be apparent.

Table of contents

Preface   vii

About the author   ix

A critical study of the search for authenticity and significance in historic wooden buildings   1

Issue of the reconstruction of wooden log cabins as a part of the interpretation of historically significant places in West and Middle Tennessee – case studies of Parkers Crossroads, Meriwether Lewis Monument and Shiloh National Military Park   13

The preliminary recognition of the condition and authenticity of historical log structures remaining in Cane Ridge community (Antioch, Tennessee)   34

Reconstruction of a group of historic wooden buildings and the authenticity of the architectural heritage structure – a case study of Wynnewood, Tennessee   77

Authenticity versus interpretation – issues of the preservation of historical wooden buildings using the example of The Tipton-Haynes Historic Site and The Historic Sam Davis Home and Plantation, significant historic farms in Tennessee.   106

Preface

Historic wooden architecture is one of the most unique types of built heritage. Erected of organic material, wood, it stays in harmony with the natural surroundings and captivates subsequent generations with its particular charm. Because of its specificity it directly depicts the richness of traditional building solutions developed in different parts of the world and at the same time it contains a record of spiritual values that were important for those by whom the architectural structures were erected.

Therefore, we should strive to ensure that this valuable part of the world`s cultural heritage is properly protected. Bearing in mind the impermanence of wood as a building material, wooden architecture requires a special approach for its preservation. It is significant that diverse cultures over the centuries have developed not only their own methods of extending the lifetime of these structures, but also, in a different way, understood the authenticity of the historical wooden building transmitted in time. And it is this last aspect, namely the issue of authenticity, which seems to be the key to capturing the character of a wooden building and the issue of its proper preservation.

This monograph discusses the issue of authenticity in the preservation and protection of historical wooden architecture from the perspective of contemporary conservation thought regarding architectural heritage. It contains five papers that show different aspects of the problem. The first text, which is also the theoretical introduction to the whole monograph, discusses the issue of understanding authenticity in relation to the adopted conservation doctrines and the consequences resulting from those doctrines for the application of the practical solutions.

Subsequent texts analyze the authenticity of the wooden architectural object using examples of specific conservation solutions and treatments implemented on selected historical log houses in the American South (the case studies come from the Appalachian region to the river lands of West Tennessee). These buildings, as important objects of architecture due to their historical and cultural values, are also structures that determine the identity of the communities in which they were erected. Conclusions and observations resulting from the examination of particular case studies are a contribution to the critical discussion of conservation methods and methodologies used in the area of western culture and their impact on the authenticity of a historical architectural structure.

The monograph particularly focuses on the considerations relating to the method of protection and conservation of wooden buildings as a form of specific historical interpretation, as well as the issue of reconstruction and translocation of a wooden historical building with regard to the level of its authenticity. In addition, analyses of the essence of historical modifications and the methods of maintaining and displaying wooden structures in relation to the requirements determining their historical and architectural authenticity are presented and discussed.

Therefore, the monograph as an important voice in the discussion on appropriate methodologies and methods of the preservation of wooden architecture, is at the same time an attempt to face the problems resulting from rapidly progressing globalization and the disappearance of local building traditions. Showing various aspects of the authenticity of a wooden building and discussing them using specific examples, it provides a rich material for further reflection and thus it is a major contribution to understanding wooden architectural heritage from a new perspective.

Prior to acceptance, the final manuscript has been reviewed by Associate Professor Klaudia Stala, Director of the Institute of History of Architecture and Monument Preservation, Faculty of Architecture, Cracow University of Technology, Poland, and by Professor Carroll Van West, Director of the Center for Historic Preservation, Middle Tennessee State University, Murfreesboro, TN, USA.

About the author

Dr Tomasz Tomaszek works as an Assistant Professor at the Department of Monuments Conservation, Faculty of Civil and Environmental Engineering and Architecture at Rzeszow University of Technology in Poland. He graduated with an MFA in Conservation and Restoration of Fine Art Objects from the Academy of Fine Arts, Krakow, Poland and an MA in Philosophy, Jagiellonian University, Krakow, Poland. He obtained his PhD (with honors) in Technical Science in Discipline of Architecture and Urbanism (specialty: history and conservation of monuments of architecture and urbanism) at Krakow University of Technology, Faculty of Architecture, Department of the History of Architecture and Conservation of Monuments, Krakow, Poland.

Dr Tomaszek furthered his professional qualifications by participating in several international specialist courses and additional studies, including:

- International Course on Wooden architecture conservation and restoration, organized by ICCROM, Petrozavodsk State University, UNESCO Chair "Wooden Architecture Research and Preservation" and Ministry of Culture of the Russian Federation; Petrozavodsk and open-air museum Kizhi, Russia;
- Course on Management and Monitoring of World Heritage Sites with special reference to China, organized by State Administration of Cultural Heritage China, ICCROM and China Academy of Cultural Heritage; The Summer Palace World Heritage Site, Beijing, China;
- International Course on Conservation of Built Heritage (CBH14), organized by ICCROM, Rome, Italy;
- 14th International Course on Wood Conservation Technology (ICWCT 2010) – A Course on the Conservation of Cultural Heritage Made of Wood, organized by Riksantikvaren - The Directorate for Cultural Heritage in Norway, Oslo, Norway; and "Architecture – Continuing Education", Norwegian University of Science and Technology, Trondheim, Norway

Dr Tomaszek is a laureate of many prestigious scholarships/fellowships, such as:

- Senior Scholar Fellowship at AATU – Tianjin University Research Institute of Architectural Design and Urban Planning, Tianjin, China; research fellowship from UNESCO/People`s Republic of China (The Great Wall) Co-Sponsored Fellowships Programme
- Senior Scholar Fellowship at Institute of Architecture, Chang`An University, Xi`An, China; research fellowship from China Scholarship Council;
- Research Fellowship at the Center for Historic Preservation, Middle Tennessee State University, Murfreesboro, Tennessee, USA; research grant from the Kosciuszko Foundation;
- Study visit in Riksantikvaren - The Directorate for Cultural Heritage in Norway, Oslo, Norway – study visit was possible due a scholarship/grant from FSS Scholarship and Training Fund;
- Scholarships (three times) "Polish Culture Worldwide Programme", Adam Mickiewicz Institute, Poland.

Dr Tomaszek has presented his research at many national and international scientific conferences as well as seminars and lectures (inter alia in Palestine, the Czech Republic, Egypt, Lithuania, Portugal, Turkey, Italy, Spain, Hungary, Russia, China, Japan and the USA), and he participated in an international research grant (Japanese-Ukrainian-Polish co-operation) "Conservation Method of Wooden Churches in Ukraine – Comparative Study of Log Construction Conservation Method between Japan and Europe"; research fund from Japan Society for the Promotion of Science (JSPS).

Dr Tomaszek is the author of several scientific papers on issues related to the preservation of historical wooden architecture, which have been published in renowned Polish and international journals.

A critical study of the search for authenticity and significance in historic wooden buildings

ABSTRACT: This paper critically discusses issues in understanding the authenticity of wooden historical buildings from the standpoint of contemporary conservation doctrine, and their implications for conservation action. Special focus is placed on analyzing where to look for significance (which is, in fact, what we intend to save) in such structures, how significance depends on authenticity and what its holders are. There is also discussion of what must be taken into account when planning conservation action to ensure that the holders of significance, and the significance itself, will be saved. Special attention is directed towards the issue of valuable contribution in the case of wooden buildings, the legitimacy of the concept of minimum intervention, an understanding of the aspect of genius loci, and the importance of aesthetic value and its authenticity in searching for significance.

Keywords: wooden historic buildings, authenticity, significance, conservation action, artistic value

1 INTRODUCTION

Where should we look for authenticity in the case of historic wooden buildings? How can we define their significance that must be saved? In trying to answer these two questions we arrive at difficult problems concerning the conservation of such structures. Keeping in mind the particular properties of wood as a building material, especially its susceptibility to rapid degradation, we face a real challenge in how to design conservation actions that will follow best professional practice and at the same time will respond to the needs of historic wooden buildings.

The concepts of authenticity (and associated with it, identity) and significance of a historic building have become central issues in contemporary theories of the conservation of architecture. Where the proper understanding of the significance of a historic building (and hence the understanding of its particular values) should help the design of a correct approach to its conservation, the search for authenticity is intended to secure an additional guarantee of the preservation of its most important aspects.

The importance of authenticity, and its central position in analysing possible alterations to the condition of the building, is very clearly marked in the essential architectural heritage policy documents. We can, as an example, recall here two of them. The first, the *Venice Charter*, highlights the responsibility of conservators to hand historic architecture to future generations *'in the full richness of [its] authenticity'* (*Venice Charter*). Whereas the second, *The Principles for the Preservation of Historic Timber Structures*, states that the primary aim of conservation is to maintain the historical authenticity and integrity of the cultural heritage. Historic buildings should therefore be preserved as authentic objects that connect the past with the present.

What then does *authentic* mean when our historic building is a wooden structure?

2 AUTHENTICITY – STATIC OR DYNAMIC VALUE?

For conservation the most important thing is an action, some kind of intervention that changes what is detrimental in the condition or context of the 'thing' in question (Bell, 2009). But in a deeper sense this action is carried out for the survival of the values of the 'thing', values which are essential for the continuity of human culture in time. '*It is not the thing per se we are trying to conserve, but its cultural significance (and so its value), that we conserve by conserving the thing*'(Bell, 2009, p.55).

The term 'thing' should be understood here in a very wide ontological sense. 'Thing' can be the name for anything that we recognize as holding significance, something tangible or material, as well as something intangible or immaterial. In particular the 'thing' is also the building, so in our considerations historical wooden architectural structure with all its tangible and intangible aspects.

Significance and value themselves are conceptual and therefore intangible, and as a result there are the following three types of conservation action (Bell, 2009):

1. Action to conserve (intangible) significance through alteration to the conditions of its tangible holders
2. Action to conserve (intangible) significance through alteration to the conditions of its intangible holders
3. Action to conserve (intangible) significance through alterations to the conditions of holders whose nature has both tangible and intangible components that are inseparable

As stated at the beginning, when planning any alteration (or conservation action) authenticity should be the main factor in the decision-making process. However, the authenticity of the built architectural monument does not relate solely to one particular area, upon which decisions about maintenance are made; on the contrary, it is a complex multi-faceted issue. Proposals laid out at *The Bergen Conference on Authenticity* (Larsen, Marstein, 1994) designated five areas (dimensions) in which authenticity can be considered essential in determining the validity of calling something a part of cultural heritage (Larsen, Marstein, 1994). They are as follows:

1. The shape or design/project (shape is something that exists, project is something intended)
2. The material/fabric or substance
3. The function or use
4. Context or place, the spirit of place (*genius loci*)
5. Techniques, traditions or technologies that include pre-industrial as well as industrial techniques and technologies

So conservation action, which has as its main goal the maintenance (or in other words, the rescue) of the significance of the wooden building, should be designed to maintain all aspects of its authenticity. But it looks like the authenticity of the wooden building –any particular dimension of it or all the dimensions together – is responsible for us seeing this building as significant, and therefore its significance depends on its authenticity. (It is actually difficult to imagine an object, for example a monument, which doesn't have or has lost the authenticity of all its aspects, and nonetheless still has cultural significance worth preserving. Maybe in this case its significance will make up for the lack of authenticity and integrity. It will probably be something completely ephemeral in all its aspects. Perhaps an object undergoing constant change? But, can we call this its significance?)

An analysis of the procedure for inscribing a historic building on the World Heritage List (Larsen, 1994) might be helpful for understanding the essence of this issue. The structure proposed for inclusion must pass (according to Article 24 of the *Operational Guidelines for the Implementation of the World Heritage Convention* UNESCO, 1988) '*the test of authenticity in design, materials, workmanship or setting*'.

Authentic in this sense does not necessarily means original (Larsen, 1994). According to Jokilehto, who quotes Article 9 of *The Operational Guidelines for the Implementation of the World Heritage Convention* (1977 revision), the evaluation of the authenticity *'... does not limit consideration to original form and structure but includes all subsequent modifications and additions over the course of time, which themselves possess artistic or historic value'* (Jokilehto, 1985).

This means that we have to accept the passage of time as the main agent responsible for constituting the dynamic state of authenticity of the building in question. It then needs to be seen as the opposite of the idealistic, static authenticity, whereby objects are frozen and removed from the context of history.

Such an understanding of the concept of authenticity is consistent with Article 11 of the *Venice Charter*. It allows us to consider, with regard to authenticity, the building and the entire history of the changes that have occurred in its construction, along with the modifications and additions resulting from repairs – the consequences of material degradation – as well as modifications carried out for aesthetic or functional reasons. However, according to the *Venice Charter*, modifications that have been executed during subsequent periods should demonstrate a *'valuable contribution'* (*Venice Charter*, Article 11). In each case, a team of specialists should rule on the question of valuable contribution proposed by the *Venice Charter*.

The concept of valuable contribution seems to be clear in the case of buildings erected of durable materials (stone or brick), however it requires far-reaching interpretation in order to be applied in relation to wooden buildings (Larsen, 1994). The changes throughout the history of buildings made of permanent materials are mainly characterized by adding new parts or layers, while in the case of wooden architecture we usually have to deal with the exchange of older elements – which are damaged or burned – for new elements.

3 THE ISSUE OF VALUABLE CONTRIBUTION

The practice of exchanging technically degraded elements for new ones, commonly acceptable in the building and conservation traditions of Far Eastern countries (Larsen, 1994), raises questions concerning the identity of the object (Jokilehto, 2006). We can assume that contemporary theories of conservation, by accepting the addition of valuable contribution, no longer take the original form or the original materials as the primary determinants of the identity of an historic building. The primary task of conservation becomes, in practice, the attempt to preserve the buildings in the state in which we inherited them, as a whole with additions acquired over the passage of time (assuming of course, as per the *Venice Charter*, that they are the *'valuable contribution'* of an earlier period). Identity therefore gains a much broader meaning and is defined as a dynamic concept, acquired throughout a building's history. Such an approach is particularly important in the analysis of the authenticity and identity of the historic wooden building, as it legitimizes recognition of identity despite the fact that a significant portion of the original, degraded wooden elements has been replaced with new elements in the intervening time. And equally, a building that retains only a residual amount of original material can be considered authentic.

It is worth noting that by accepting the concept of valuable contribution we agree, at the same time, to recognize that the beauty (and the aesthetic value) of the building comes also from the patina of time (Larsen, 1994). This patina has inherent value as a document, and so it also gains historical value. The preservation of that patina is a necessary prerequisite for maintaining the historicity of the building.

There is the danger that acceptance of the acquired historical values of the building as foremost could lead to the loss of its original aesthetic quality. Especially if the secondary additions carried out during the history of the building weaken its original aesthetic. Therefore it is debatable whether the recommendation to preserve the valuable contributions of each period (in Article 11 of the *Venice Charter*), which forms the criteria for assessing the authenticity of the building, can be unambiguously defined and enacted on the ground.

The roots of the theories that promote the acceptance and the preservation of the building in the condition in which it reached our time (with the modifications executed throughout its history) can already be seen in the first half of the nineteenth century (Jokilehto, 1986). In 1839, A.N. Didron noted that:

'it is better to preserve than to restore and better to restore than to reconstruct' (Murtagh, 1990).

This type of approach to conservation can be defined as minimum intervention. The method of minimum intervention – considered by its proponents as the most appropriate – gained supporters at the end of the nineteenth century (Jokilehto, 1986). Its most important follower was John Ruskin and it has been adhered to until today by some members of *The Society for the Protection of Ancient Buildings* in Great Britain (Jokilehto, 1986).

Such an approach to conservation was developed theoretically in the paper released by Alois Riegl in 1903, where the concept of age value was introduced. (Riegl, [1903] 1929). This concept requires us

'...to assume that everything which history has changed is irreversible and as such has become part of the monument' (Neuwirth, 1987).

As long as the structural whole and stability of the historic building is intact, the implementation of a policy of so-called minimum intervention seems to be correct and remains in accordance with the requirements of contemporary concepts of conservation of architecture. This method, similar to the concept of valuable contribution works without major problems in the conservation of buildings erected from durable materials (such as brick or stone). In these cases the original material and the subsequent valuable additions can be safely preserved without compromising the building's global structural stability.

The matter is different when we consider the issues of conservation relating to structures built of wood. Bearing in mind the characteristics of wood as a building material and its rapid process of degradation, the theory of minimal intervention must be specially interpreted (Larsen, 1994). Therefore, as we see in practice, minimum intervention often results in the exchange of a large amount of elements which – as completely degraded components – need to be replaced by new ones. Thereby the carrying out of minimal intervention takes the form of far-reaching conservational interference which is the only guarantee of re-establishing the building's threatened stability.

Does the wooden building therefore retain its cultural significance even though a considerable amount of elements has been exchanged? Will the cultural significance of the wooden building be the same (its nature unchanged) when we accept its authenticity even though the fabric is not original? Where should one look for significance then, and what are the supposed carriers of significance?

4   RECOGNITION OF SIGNIFICANCE

According to Bell, there are two components necessary to define an act of conservation: (1) the holder of significance (the 'thing'), and (2) the particular significance it (the 'thing') holds (Bell, 2009). It then follows that the right conservation action is based on the full understanding of both.

As for the first component, the analysis of the nature, condition and context of a 'thing' (in our case, an architectural structure) is generally well understood by professionals (such as architects, conservators or technical experts), as well as the implications of their analyses. But there is some ambiguity about the second component, a component that seems to be foremost when defining action (Bell, 2009). It becomes especially problematic to understand the implications of the conclusions collected from the observation of the particular significance of the particular building correctly for the planned conservation action.

Following on from this understanding, in the case of wooden buildings, it is quite easy for conservators to describe the particular nature, form or condition of a particular structure (building). In the main, there are also few problems in discussing conservation issues concerning its fabric – wood – which, as an organic material, is very sensitive and undergoes rapid degradation. But the majority of this concerns the tangible aspects of the building (let us say the holders of significance), aspects that are very important for safeguarding the authenticity of the building, but not in themselves sufficient for understanding the significance of the wooden structure in question. (Of course this problem applies to any kind of built heritage, not only wooden objects. But in the case of wood, when it is used as a basic material for the erection of an architectural structure, its ephemeral character is more evident and thus the significance of this kind of structure cannot rely solely on an understanding of authenticity based on the original fabric alone).

So now we face the problem of the proper assessment of the significance which the particular historic wooden building holds. Keeping in mind what we have said on this point, the importance of understanding the authenticity of the wooden structure, in which case authenticity can be recognized as a dynamic process should again be stressed. This means, then, that the significance of the wooden building will be more attached to its ephemeral nature, which requires the process of repeating or redoing as a necessary condition for its existence in time. And thus significance will be kept only when this process of redoing or repeating (the continuous process of exchanging degraded elements for new ones) is maintained. Thanks to the continuous replacement of elements, the original craftsmanship skills necessary to produce new elements in a traditional yet original way will be passed on. By keeping these skills alive, the continuing authenticity of wooden structure is assured.

Going back to significance, when we postulate the dynamic nature of authenticity, we might notice that the dynamic process of repeating is included in our understanding of the significance of the wooden building. But this repeating doesn't only concern the fabric that – as we noted – must be exchanged from time to time, but also concerns the aesthetic value of the structure. Thus repeating is constantly required to keep the aesthetic value unchanged, because the wooden structure degrades very rapidly compared to structures erected from permanent material.

This seems particularly crucial in the case of wooden religious buildings, as the aesthetic completeness of the temple was one of the main tasks of its keepers (Larsen, 1994). However, the life of a wooden place of worship in some traditions can be considered an ephemeral phenomenon (Elleh, 1996) and thus their material (physical) form is intended only to fulfil their specific sacral function in a given time period. The best example of this may be Shinto temples in Japan, which are destroyed and rebuilt in the same material form every 20 years. This periodic destruction and rebuilding of the shrine symbolizes the life and death cycle with periodic renewal, whereas wood – the building material, considered to be sacred and looked upon as a symbol of life (Henrichsen, 2004), (Enders, 1999). If we refer to the essential archetypical concept of the returning order of things in time, outlined in the *Myth of the Eternal Return* (Eliade, 1974), it will be obvious to accept that any established form shaped of the material substrate has its beginning and end, although after its disappearance the cycle repeats and the new beginning and then the new end of the same established form happens, while the material substrate – or the combinations of its arrangement – is actually different. And this cycle is infinitely repeated. Thus form has an eternal existence, while the material substrate used for achieving (or building) this form undergoes constant change. This argument initially seems to be applicable to any kind of built structure, however, we need to note that permanent materials were used by builders in the hope of erecting monuments which would last forever (for example the pyramids in Egypt), while in the case of wooden structures they most likely had a complete understanding of their temporary nature. This character trait of wooden buildings is directly connected to the nature of wood. Wood itself – especially species coming from moderate or cold climates, where wooden architecture is actually the most diffuse – as a living organism repeats the annual cycle of recovery and death, year after year. And finally each plant, similar to any living organism, has an intangible genetic, specific form that has continuity in time and guarantees the continued existence of this tree in time, even though the particular single tree's cellular body has a limited lifetime. Thus we should definitely regard wood as a living material for building and accept all the consequences that arise from this.

So, what will be more important for a conservation action applied to a wooden building – to understand the nature of the holder of significance and to try to maintain its authenticity (even if the fabric is not original) or to strive to understand the particular cultural significance that this particular building holds and look for the authenticity upon which it depends? Well, as we already have noted above, both of these aspects are equally important and understanding both of them is necessary to allow a correct definition of the act of conservation.

Assuming again that, as conservators, we have better insight, or more knowledge, about the nature of the wooden heritage building as a holder of significance, let us focus in more detail on the significance it holds. As noted above, significance refers to values that are exceptionally important, due to which this particular structure is worth saving for future generations. But what kind of values are these? .In the case of any historic building, including wooden ones, we can talk about its historic values. These are usually mostly connected to the tangible aspects of the site (building) – like for example the historic (or documental) fabric, but they can also be related to intangible aspects that are of historical importance. They can be, to give an example, the particular traditions of a place, like local beliefs or rites.

But there is one kind of value, let us call it the artistic value or aesthetic value, that cannot be captured by (or when) analyzing or understanding or reading properly only the tangible aspects of the historic building (Bell, 2009). By describing all of the tangible aspects, let's say, for example, the shape, colour, structure (et cetera) of the wooden building we are not able to define the artistic value itself (artistic value in itself is not quantifiable). The artistic value eludes any of these ways of capturing it, is beyond fabric, beyond tangible, beyond something which can be easy measured and classified. (By describing and comparing all the tangible aspects of different objects (sites) we are not able to define which of them has a higher artistic value. The artistic value is beyond this simple way of describing and comparing elements). But, even if not tangible, it is something that heavily influences our particular interest in any particular object.

5   ARTISTIC (OR AESTHETIC) VALUE

Artistic (or in other words, aesthetic) value is the basis for recognition of the aesthetic significance of the building. An observation of the aesthetic significance of the site is then directed towards appreciation of its aesthetic value.

The aesthetic, or artistic, significance seems to be one of the most important aspects in analyzing the significance in general of any kind of object, including historic buildings. The nature of aesthetic significance is of course intangible, but what will be the nature of its holder? Is the holder of aesthetic significance tangible or intangible? Our initial response tells us that obviously the holder must be tangible as the form, colour, shape, structure and smell of the fabric are responsible for bringing to life the aesthetic value and thence the aesthetic significance. Even though, as noted before, we are not able to define an artistic experience by measuring or comparing all the individual tangible aspects of the site, we nevertheless feel that because of them and their arrangement we are experiencing a work of art at this particular site and at the same time enjoying the aesthetic significance of the site. But there is also a more considered response, something beyond the initial one, something that seems to go further and point at the intangible or elusive aspect. There is something telling us that there must also be some 'beyond-tangible' holder of the artistic significance, something that we might call the 'hidden spirit' of the site. (Can we then regard the *genius loci* of the site as an intangible aspect and at the same time also as the holder of its artistic significance?)

6   AESTHETIC SIGNIFICANCE, ITS AUTHENTICITY AND WOODEN BUILDINGS

For now let's return to the assumption that aesthetic significance relies in some important ways on all the tangible aspects of the site (in this case, building). So, how can we try to define an action to conserve the intangible artistic significance that rests on the tangible aspects of the building?

The issue of the significance of buildings seems to have been problematic from the beginnings of the conservation movement. According to Bell, already in the Eighteenth century there was a clear dividing line placed between historic monuments (the important remnants of civilization that must be saved in the condition in which they have survived into our time) and architecture – valued for its use and appearance (Bell, 2009). In other words, and using Cloquet, for this division we can call the first group dead monuments and the second group living monuments (Cloquet, 1922). Among the second group, the buildings chosen because of their exceptional visual (artistic or aesthetic) values have been also destined for saving. But this artistic significance, unlike in the case of historic monuments, was seen as something that needed to be kept by treating them as any other kind of art object. They had to be maintained in the best condition possible and retouched to show them in their full glory. In other words, all the tangible aspects of the object must be restored to their initial appearance.

This paradigm for approaching the built environment left its imprint on the conservation practice of the following decades. While the significance of the building was seen in relation mainly to its artistic value, this was a reason for the acceptance of reproduction, replacement and even stylistic completion. Because, as it was understood, in the case of buildings where the artistic significance took priority, they had to be presented as complete works of art in their entirety.

The question that follows is whether this aesthetic or artistic value must be authentic in order to be significant and what it will mean or what its consequences are. Keeping in mind the wooden structures, which this paper considers, the idea of restoring them constantly over time to their original appearance due to the necessity of maintaining the authentic artistic value seems to be the obvious response. If we agree that the artistic value is attached to the tangible aspects of the building then, because wood degrades very quickly, the replacing of the original material with new elements will fulfil this maintenance objective. But does the copy have the same artistic value as the original? As stressed before, we can`t define the artistic value only by measuring or analyzing the tangible aspects. So, perhaps we can state that it actually does not matter if it is a copy or original as long we can read the artistic value and find its significance. And, following on from this, we should re-think the idea of authenticity, as artistic value cannot be easily contemplated at all within this category. What is an authentic artistic value? Can something like this be experienced under any circumstances?

From a theoretical or philosophical point of view we are not able to find an ultimate definition of art and thereby artistic value. Is it something that is felt subjectively or does it have an objective dimension? Or maybe it has both of these aspects at the same time? These are historic philosophical problems that have not been resolved to this day. This issue was also discussed by some of the theoreticians of architectural conservation, among whom it is worth mentioning Alois Riegl, who in 1928 contended that, because the concept of artistic value was unworkable, it should be abandoned or replaced by its more evidential aspects (Riegl, 1928).

To make matters even more complicated we cannot be sure whether we can think about art as a concept that exists permanently (it would have been understood in the past as an 'Ideal Object' existing (or located) in Plato's world of ideas). From our experience in our time we are convinced that art is perceived differently in other historical eras. To quote Bell, who refers to Riegl, we can assume that art – *'whatever it is – can be appreciated by contemporary eyes only in their own contemporary way, and never in the way of its original creators', so it [art] has a "present-day' (as opposed to historic) value'. … To the 'aesthetic-ally educated modern person' he [Riegl – comments of author] assigned a heightened appreci-ation that could manage to see significance in art of a different era, though this would not necessarily be to the taste of his or her own age, and which modern artists, bound by their time, could never achieve'* (Bell, 2009, p. 58).

So, if we agree with this point that art has only a present day value, the problem of the authenticity of the art of the wooden building becomes more complicated. If we again refer to Riegl, and cite his remark that generally *'only the new and whole things tend to be con-sidered beautiful; the old, fragmentary, and faded are thought to be ugly'* (Riegl, 1928, p. 56) we will be in some way describing the way of understanding this issue in the traditions for the conservation of wooden architecture in Far Eastern countries, among them Japan (Larsen,1994). The concept of beauty, and thus art, seen in the building which is incom-plete, ruined or aesthetically changed by patina seems to be most developed in the Western world (Jokilehto, 2004), but even there it is rarely applicable in the case of wooden struc-tures. Even if we accept a certain 'patina of time' on wooden architectural structures, they are almost never left in a degraded state for purely aesthetic reasons.

So what is so special about wooden buildings? It looks like our present day concept of the artistic value associated with wooden buildings is very similar to that which was under-stood in earlier historic periods, and that we have a different understanding of the holder of this value than in the case of buildings erected from permanent materials, which – as noted before – may be appreciated even if they are ruined. Firstly, we regard the wooden building more as a tradition of building or a process, so the artistic value will be directly attached to something that is dynamic in nature. It is akin to appreciating something as being ephemeral while at the same time, by repeating or redoing its tangible aspects, chan-ging it into something permanent. This is inherent in the nature of wood – to be ephemeral and permanent at the same time. So, the artistic value, and the resulting artistic significance of a wooden building, does not rely simply on tangible aspects that must be repeated in the same order to keep this value (and significance) unchanged. Rather, it is something more like an appreciation of the unique character of the permanent shape clothed in an ephem-eral fabric. Of course it is beyond dispute that artistic value can be observed only as long as the tangible aspects of the building are present, but this value is not simply related only to the authenticity of those tangible aspects. With minimal discussion we may agree that artis-tic value is not limited to the authenticity of the fabric (which contains mostly the evidential value) and this assertion then has essential meaning for conservation action. (The most common solution adopted in the conservation approach for decades has been to group the art-related but primarily evidential aspects of value under the umbrella term of artistic value. All these aspects were conserved by the same approach as for any other type of evi-dence but, as has been argued above, that does not necessarily protect the quality of the art itself, or its significance).

To guarantee the significance of the wooden building we need to maintain not only the whole set of holders of significance that retain authenticity (like place, shape, etc.), but most of all its observed artistic significance. But this artistic significance, directly resulting from the artistic value of the building, will always require for its ongoing existence the pro-cess of repeating or re-doing necessary for the continuity of a wooden structure. To keep this value alive, the maintenance of the intangible tradition is of primary importance, in addition to the shape or character of a place. And even though the artistic value will always be a contemporary value, this process of maintaining it shows its real authenticity, in that it is a never-ending process. In other words, we can say that the wooden building is a process and inside it is its significance, which can be observed explicitly through the artistic value found each time that materialization of the process is carried out. It is more important to safeguard the art of the building rather than its fabric, and this art is included in its formation and disappearance. And because of all of this the artistic value actually becomes the historical value due to its holding of the permanent evidence of the past. So the historical evidence will not be kept by the fabric but instead by an aesthetic value transferred over time. This aesthetic value can then be called an historic document. (We can call this a conservation approach that aims to treat the aesthetic aspect as historical evidence).

If we eliminate artistic value from our discussion of significance the only thing left will be architectural technical knowledge about the building and materials. Designing the conservation action upon this knowledge only will never allow the transmission of significance and may result in inappropriate alteration. So the question is more how far an object, in our case a wooden building, can accept change, and what sort of change it can accept, without losing cultural significance of any kind. The simplest answer is in the point that the real cultural significance here is in the continuance of the wooden building tradition. And authenticity is in the process, rather than in the material or fabric. All the tangible aspects are only attributes that are attached to core essence of the 'significance of process'. So their importance is not in their unchanged existence but in their acceptance of change.

And finally, equally important is the context of the building, the social, geographical, environmental or natural context of the place and its surroundings. The constant process of change that takes place here is a part of the life of a wooden building and is also the pattern for maintaining the authenticity of the artistic value of the place and the material form of the wooden structure. So, for a correct conservation action this aspect should also be properly examined and the understanding of the significance must also include a full understanding of the context. Any change in this context could be a reason for an alteration in significance. So, the conservation action cannot be planned separately, focusing only on the tangible aspects of the wooden building. It must regard 'the whole or oneness' and not only its parts. (The compositional entity or 'oneness' of the site (building) is not simply the sum or result of the arrangement of all individual elements. Referring to Cesare Brandi, 'oneness' can be understood as the unique quality of a particular aesthetic experience, its qualitative not quantitative value, and, above all, the absolute necessity of designing any conservation action from a vision of the work (the site), as an entity, a whole, not as a sum of parts. (Brandi, 1963).

To conclude these considerations of artistic value and artistic significance it is worth remembering the highly influential Australia ICOMOS Burra Charter, published in 1979 (revised 1999). There, probably for the first time, important stress was placed on sites' (or buildings') full sensory impact, intentionally or unintentionally created. Aesthetic (or artistic) value was defined as including *'aspects of sensory perception for which criteria can and should be stated. Such criteria may include consideration of the form, scale, colour, texture and material of the fabric; the smell and sounds associated with the place and its use'* (Guidelines to the *Burra Charter*: Cultural Significance; 2.2). However the most important task of this document was to draw attention not only to the significance of a site's visual impact, but to its effect on all the other senses. And once artistic value has been acknowledged we must face the problem how to recognize it and protect it in practice. So, let us close this paragraph referring again to Cesare Brandi, who claimed that the essence of conservation is *'the methodological moment in which the work of art is recognized, in its physical being, and in its dual aesthetic and historical nature, in view of its transmission to the future'* (Brandi, 2005, p. 47).

7   *GENIUS LOCI* AND AUTHENTICITY

Let's return to what was noted at the beginning that authenticity can be looked at from different perspectives and that the one of the most essential elements that validates the inclusion of something as cultural heritage is the authenticity of the spirit of place (*genius loci*). Now we have to ask, what is the nature of the spirit of place? How can we define its significance, the significance which we are trying to preserve as authentic and, even more importantly, what are the holders of this significance?

According to Petzet it would be highly improper to think in so-called *'dialectics between spirit and place, the intangible and tangible'* (Petzet, 2009, p. 64) because it simply equates *"spirit' with 'intangible/immaterial' and 'place' with 'tangible/material"* (Petzet, 2009, p. 64). It must be noted that the place may also have intangible dimensions, and can also be *'an ideal or unreal, at any rate an intangible place'* (Petzet, 2009, p. 64). The tangible and intangible values are not then separate, rather, they are (according to a very helpful definition by Mounir Bouchenaki [after Petzet]) *'two sides of one coin'* (Petzet, 2009, p. 65) which, thanks to the particular *genius loci* exist as a natural unity. So the holder of significance of the particular *genius loci* of the building can then be seen as corresponding to the individual atmosphere. This atmosphere, or aura, may be the product of tangible as well as intangible aspects, and its existence is dependent on the particular combination the both of them found in situ. For obvious reasons it cannot be transferred to any other place.

To better understand the nature of *genius loci* we may assume, according to the definition in a late-classical commentary on Cicero, that monuments *'should evoke remembrance of something'* [*omnia monumenta sunt, quae faciunt alicuius rei recordationem*] (Petzet, 2003). This simply means that some spiritual message, let's say an intangible idea, is the reason for its existence. So the monument is in every case more than an architectural structure built of a certain material, it is actually an idea that has taken on a particular shape. The fabric from which the monument is erected, as an object of remembrance, can thus be *'just as variable as the degree of 'materialization' of the spiritual message that the monument represents'* (Petzet, 2009, p. 65). This applies especially to historic wooden buildings, erected from this fabric that is so ephemeral. These buildings are in need of renewal again and again, and even *'the mere replica of a monument that no longer exists materially could still 'evoke remembrance of something''* (Petzet, 2009, p. 65). So, in the case of an historic wooden building the intangible message or idea that takes this particular shape is more like the permanent form that has its continuity guaranteed by the dynamic process of redoing/repeating. And this process, as we have already said, keeps this idea alive and is actually a prerequisite for its existence. The permanent form we have mentioned doesn't apply, however, only to the shape of the wooden architectural structure, it is also applicable to what can be understood as the context of the place. This permanent form would then, as its message, shape the cultural landscape surrounding this particular place, as well as the existence of the local culture and the knowledge of local building traditions. It can be seen very clearly in the influence of the landscape on wooden architecture and at the same time the influence of wooden architecture on the landscape. This permanent form is then responsible for the aura linked to a particular place and embedded in history, and this aura remains as long as the permanent form is not forgotten, even when the wooden monument is hardly comprehensible as historic fabric.

So, under these circumstances, the significance of the particular *genius loci* is an important consideration as far as questions of how to conserve, restore or renovate are concerned. We must ask ourselves, then, if our planned conservation action is correct for the individual *genius loci* of the wooden building in question and whether we are preserving the *'spiritual message of a monument which, compared with a long history, has been entrusted to us only for a short time'* (Petzet, 2009, p. 67). Keeping all of this in mind we must remember that the *genius loci* of the wooden building is strictly bound to the permanent form of the spiritual message communicated in this building. So the permanent form, which depends on the dynamic process of redoing/repeating, here takes on the role of the holder of its significance, and, only by ensuring the continuity of this permanent form can we guarantee the authenticity of the particular spirit of place of the historic wooden architectural structure.

8   FINAL REMARKS

Now, following the findings of the *Yamato Declaration on Integrated Approaches for Safeguarding Tangible and Intangible Cultural Heritage* we are more aware than ever of the importance of safeguarding intangible cultural heritage. But we still do not appreciate sufficiently that intangible aspects remain present in all historic buildings, even those which have been abandoned by communities that have lost their cultural continuity. And those intangible aspects are crucial for ensuring the authenticity of this architectural heritage. Whether we are able to understand easily these intangible aspects or whether they are elusive to us, we must understand their precedence in the design of any conservation approach/action.

This applies most significantly to historic wooden buildings, as those structures are ephemeral by their nature. Their tangible aspects are only the materialization of the spiritual message and they will keep their authentic significance as long as their intangible features continue to speak to us in the proper way. The conservation of historic timber structures can thus be understood as a way to allow them – by accepting the understanding of authenticity in a broad sense – to express themselves in their own historical language. This means that we are not only passing on historic buildings to future generations, but also ensuring that they will speak to them in an authentic language.

REFERENCES

Australia ICOMOS. 1999. *Burra Charter, Australia Charter for the Conservation of Places of Cultural Interest*. Burra: Australia ICOMOS Incorporated. 1.4.

Bell D. 2009. *The naming of parts*. In *Conserving the authentic: essays in honour of Jukka Jokilehto*; Editors: Stanley-Price N., King J.; Rome: ICCROM Conservation Studies 10, 55–62.

Brandi C. 1963. *Teoria del Restauro*. Roma: Edizioni di Storia e Letteratura.

Brandi C. 2005. *Theory of restoration*. Firenze: Nardini.

Cloquet L. 1922. *Traité d'architecture: éléments de l'architecture; types d'édifices; esthétique, composition et pratique de l'architecture*. Paris: Liége: C. Béranger.

Eliade M. 1974. *The Myth of the Eternal Return or Cosmos and History*. Princeton: Bollingen Series XLVI, Princeton University Press.

Elleh N. 1996. *African Architecture: Evolution and Transformation*, New York: McGraw-Hill.

Enders S., Gutschow N. (ed.). 1999. *Hozon: Architectual and Urban Conservation in Japan*, Stuttgart/London: Edition Axel Menges.

Feilden B. M. 1984. *A Possible Ethic for the Conservation of Timber Structures*. Printed as an appendix in: Charles F.W.B. *The Conservation of Timber Buildings*.

Henrichsen Ch. (Author), Bauer R. (Photographer). 2004. *Japan Culture of Wood: Buildings, Objects, Techniques*, Basel: Birkhauser.

ICOMOS. 1964. *International Charter for the Conservation and Restoration of Monuments and Sites (Venice charter)* [online] [accessed 29 June 2014]. Available at: <http://www.icomos.org/charters/venice_e.pdf>.

ICOMOS. 1994. *Nara Document on Authenticity* [online] [accessed 15 September 2014]. Available at: <http://www.icomos.org/charters/nara-e.pdf>.

Ito, N. 1995. 'Authenticity' Inherent in Cultural Heritage in Asia and Japan. In: K.E. Larsen & N. Marstein, eds. *Proceedings of the Nara Conference on Authenticity in Relation to the World Heritage Convention*. Trondheim: Tapir, pp. 35–45.

Jokilehto J. 1985. *Authenticity in Restoration Principles and Practices*. APT Bulletin 17, nr. 3-4: 5-11.

Jokilehto J. 1986. *A History of Architectural Conservation. The Contribution of English, French, German and Italian Thought towards an International Approach to the Conservation of Cultural Property*. PhD diss., The University of York. The Institute of Advanced Architectural Studies.

Jokilehto J. 2004. *A History of Architectural Conservation*. Oxford: Elsevier Butterworth-Heinemann. First published Oxford: Butterworth-Heinemann.1999

Jokilehto J. 2006. *Considerations on authenticity and integrity in world heritage context*. City & Time 2 (1): 1. [online] URL:http://www.ct.ceci-br.org

Larsen K. E. .1994. *Architectural Preservation in Japan*, ICOMOS International Committee, Trondheim: Tapir Publishers.

Larsen K. E., Marstein N. 2000. *Conservation of Historic Timber Structures. An ecological approach.* Butterworth-Heinemann Series in Conservation and Museology, Reed Educational and Professional Publishing Ltd., Great Britain.

Larsen K.E, Marstein N. 1994. *Conference on Authenticity in Relation to the World Heritage Convention. Preparatory Workshop. Bergen, Norway, 31 January – 2 February 1994. Workshop Proceedings*, Trondheim: Tapir Publishers.

Larsen, K. E. (ed.), 1995. *Nara conference on authenticity in relation to the World Heritage Convention*, Proceedings, Nara, Japan, 1-6 November 1994, UNESCO, Agency for Cultural Affairs of Japan, ICCROM, and ICOMOS, Trondheim, 77-99.

Murtagh W.J. 1990. *Keeping Time. The History and Theory of Preservation in America*, New York: Sterling Publishing Co., Inc.

Neuwirth F. 1987. *Values of a Monument in a New World*. In *ICOMOS 8th General Assembly and International Symposium*. Symposium Papers 1. Washington D.C.

Operational Guidelines for the Implementation of the World Heritage Convention. 1988. UNESCO, article 24.

Petzet M. 2003. *Place – Memory –Meaning: Preserving Intangible Values in Monuments and Sites*. Introductory lecture to 14th general assembly of ICOMOS 'Place - Memory - Meaning: Preserving Intangible Values in Monuments and Sites', Victoria Falls, Zimbabwe, 2003; http://www.international.icomos.org/victoriafalls2003/index.html (4/4/2015).

Petzet M. 2009. *Genius loci – the spirit of monuments and sites*. In *Conserving the authentic: essays in honour of Jukka Jokilehto*. Editors: Stanley-Price N., King J., Rome: ICCROM Conservation Studies 10, 63-68.

Riegl A. [1903] 1929. *Der moderne Denkmalkultus. Sein Wesen und seine Entstehung*. Reprinted in *Gesammelte Aufsatze*. Augsburg-Wien.

Riegl A. 1928. *The Modern Cult of Monuments: Its Character and its Origin* In *The nineteenth-century visual culture reader*. 2004. Edited by Schwartz V., R. and. Przyblyski J., M. New York: Routledge.

The Principles for the Preservation of Historic Timber Structures. 1999. Adopted by ICOMOS at the 12th General Assembly in Mexico, October 1999. Available at: http://www.international.icomos.org/charters/wood_e.pdf, 29/06/14.

Yamato Declaration on Integrated Approaches for Safeguarding Tangible and Intangible Cultural Heritage. Adopted on International Conference on The Safeguarding of Tangible and Intangible Heritage Organized by the Japanese Agency for Cultural Affairs and UNESCO 20-23 October 2004, Nara, Japan.

Issue of the reconstruction of wooden log cabins as a part of the interpretation of historically significant places in West and Middle Tennessee – case studies of Parkers Crossroads, Meriwether Lewis Monument and Shiloh National Military Park

ABSTRACT: Reconstruction of wooden historical vernacular buildings as a part of the interpretation and preservation of nationally important historical places from the perspective of maintaining the authenticity of these architectural structures seems to be particularly problematic. The paper discuses this issue using examples of log cabins located in two geographic regions of the U.S. state of Tennessee – its Middle and West parts of Grand Division. The study explores three significant historical wooden structures in particular: McPeake Cabin in Parkers Crossroads, the Cabin at Meriwether Lewis Monument in Natchez Trace Parkway and the William Manse George Cabin in Shiloh National Military Park. All of these buildings are aimed at re-creating and reflecting a specific moment in history as well as at showing the original character and the "authentic" context of the place. Therefore they represent conservational approaches that by fulfilling these objectives face the problems of falsification and the question of the authenticity of these historical buildings.

1   INTRODUCTION

The issue of maintaining the authenticity of wooden heritage buildings takes a very special place in the discussion of the preservation of architectural heritage. The validity of relocating historical wooden structures to new sites, thus into new spatial and semantic contexts, seems to be especially problematic. When this action is implemented in order to introduce information resulting from the interpretation of the site of an important event in national history, the procedures taken should seek to avoid distorting the original historical narrative.

This article critically examines the issue of reconstruction of wooden log cabins as a part of interpretation and preservation of nationally important historical places in two geographic regions of the U.S. state of Tennessee – the Middle and West parts of the Grand Division. Therefore, three significant case studies of preservation are presented and explored, and they are as follow: the McPeake Cabin in Parkers Crossroads (West Tennessee), the Cabin at Meriwether Lewis Monument in Natchez Trace Parkway (Middle Tennessee) and the William Manse George Cabin in Shiloh National Military Park (West Tennessee).

All of these buildings are directly related to the important historical events that took place in Tennessee. Thus, they are connected not only by a similar intensity of historical significance, but also by a certain type of symbolism that shapes the national consciousness of the American people. For this reason, their existence as historical monuments has a special dimension, and the conservation solutions used are an attempt to reflect the character of the place and consolidate important events in the history of the American nation.

2   MCPEAKE CABIN IN PARKERS CROSSROADS

McPeake Cabin in Parkers Crossroads (Historical Marker HMDJ3) is located in Wildersville, Henderson County, TN (Photo 1). It was built in 1851 by Robert and Permelia McPeake, originally standing near Rock Hill, Tennessee. In 2006 Danny and Rose Gardner donated the cabin to the Parker's Crossroads Battlefield Association. After painstakingly recording the structure, it was then dismantled and moved to the battlefield. After reconstruction (reassembling), the cabin was dedicated in May 2008.

Photo 1.   McPeake Cabin in Parkers Crossroads (photo T. Tomaszek, 2018).

According to the historical sources the McPeake family came to Middle Tennessee from Pennsylvania in the mid-19th century [Historical Marker Project, 2018], and some of the family members settled in Henderson County. Robert Carroll McPeake, was living with his brother a few miles east of Lexington, about 10 miles south of current location of the cabin. In 1850 or 1851, Robert married Permelia Reed, granddaughter of Joseph Reed, the first permanent settler in Henderson County. The couple purchased land near Rock Hill, about five miles east of Lexington, where they built their house. The log cabin was erected on the double-pen (or two rooms) floor plan and in the type of building known as "dogtrot"[i] [Jordan, 1985]. It was a typical dwelling for this time in Tennessee. In all probability McPeakes built their dwelling from the trees cut on their land[ii] (Photo 2).

Robert and Permelia lived there for many years, raising twelve children in the cabin. The building remained in the McPeake family for generations and many descendants enjoyed this cabin as their home.

The Battle of Parker's Crossroads occurred on 31st December, 1862. It was part of Confederate Brig. Gen. Nathan Bedford Forrest's cavalry raid into West Tennessee and at Parker's Crossroads he was attacking Colonel Cyrus Dunham's Union brigade when he was surprised by a second Union brigade, commanded by Colonel John Fuller, attacking him from behind. Forrest's troops pushed back Fuller and moved around Dunham's

---

i. A dogtrot is a one-story house on the double-pen floor plan, with two rooms (or pens) separated by a passage, the whole covered by a gable roof. Typically it has two chimneys, located on the both sides of the building and on the exterior walls (Jordan T. G. 1985. *American Log Buildings. An Old World Heritage*, The University of North Carolina Press, Chapell Hill, p. 24)

ii. Homes similar to the McPeake cabin were erected in large numbers in Tennessee in the 18th and 19th centuries. Log houses were fairly easy to build as timber was abundant and processed lumber was very expensive. In fact, many surviving 18th and early 19th century historic homes have a log structure at their core, covered later by remodeling or the addition of weatherboard cladding.

Photo 2. McPeake cabin before its transfer to Parkers Crossroads (photo Brandon McPeake, source: Genealogy Trails History Group. 2019. The McPeake Heritage. http://genealogytrails.com/tenn/henderson/mcpeakecrossroads3.html).

already battered forces to reach and cross the Tennessee River [The American Battlefield Protection Program, 2015]. Many call Parker's Crossroads one of Nathan Bedford Forrest's key battles [The American Battlefield Trust, 2019].

There is strong evidence that there were several log cabins at Parker`s Crossroads at the time of the battle [Historical Marker Project, 2018]. There is even evidence that one cabin stood near the McPeake Cabin's current location (on the opposite side of the old Lexington-Huntingdon Road) [Historical Marker Project, 2018]. According to the historical sources this cabin was most likely known as "Parker House", the dwelling of the Parker family living in this area. This was the site of the battle's turning point as "while Forrest was parlaying with Dunham for the surrender of the Union Brigade, Colonel Fuller`s entire Ohio Brigade arrived from the north behind the Parker House" [The Parker's Crossroads Battlefield Association, 2000-2012].

Photo 3. Parker`s Crossroads Battlefield (photo T. Tomaszek, 2018).

The first attempts to preserve Parker`s Crossroads Battlefield took place in the early 1980`s [Parker's Crossroads, 2016]. In the following years, thanks to many associations (primarily the Civil War Trust in a consort with its partners, including the local group The Parker`s Crossroads Battlefield Association) a significant amount of the battlefield was purchased [Wikipedia…, 2019, Battle of Parker's….]. In 2001, thanks to the Tennessee Wars Commission, the battlefield was designated on the National Register of Historic Places [Parker's Crossroads, 2016] (Photo 3). Since then, much effort has gone into interpreting the site. One plan was the transfer of a building from a similar period and with a similar family history to the place where "Parker House" originally stood, as nothing from the original Parker House is preserved. And that`s how the public interpretation of the "Parker Family project" began. This operation was aimed at recreating the original character of the site and showing an "authentic" context as it might have appeared during the battle. The "reconstruction" of the site's significant architectural object reflects a specific moment in history. However, this reconstruction did not rely on the construction of just any architectural shape, but on a wooden structure that was architecturally, semantically and historically similar to the original lost house.

And so, a wooden building was moved to Battlefield, a building that once belonged to another family with a similar history living near Parker`s House. At its new location the building was modified for the purpose of obtaining a specific architectural form: a pioneer home (cabin) in Tennessee during the Civil War era. To achieve this, the roof was extended from the front and from the back of the building, creating typical arcaded spaces (porches), and the chimneys were added on the sides (Photo 4, 5).

Photo 4. McPeake cabin during its reassembly at Parkers Crossroads (photo Brandon McPeake, source: Genealogy Trails History Group. 2019. The McPeake Heritage. http://genealogytrails.com/tenn/henderson/mcpeake crossroads3.html).

Photo 5. McPeake cabin during its reassembly at Parkers Crossroads (photo Brandon McPeake, source: Genealogy Trails History Group. 2019. The McPeake Heritage. http://genealogytrails. com/tenn/henderson/mcpeakecrossroads3.html).

Although the change in the form of the roof itself is not evident at first sight, the construction of chimneys may raise objections because they give the impression of some kind of new addition. Even though they refer to the original shape of the chimneys at "Parker House," they are constructed in a "modern way" and not following traditional solutions [Weslager, 1969, p. 15], [Rehder, 2012, p.23][iii], (Photo 6).

Despite the fact that during the process of reassembly original woodworking technologies were used (Photo 7) and the building was embedded in a traditional way directly on loose fieldstones[iv] [Weslager, 1969, p. 15], (Photo 8), a secondary material[v], Portland Cement (Photo 9), was used as chinking between the logs forming the walls. This type of non-traditional, contemporary technological solution (chinking made of Portland Cement) was typically used for restoration of wooden buildings in the south-eastern USA in the second part of 20th century.

iii. See the information about traditional chimneys in the part of this paper discussing The William Manse George Cabin in Shiloh National Military Park

iv. According to Weslager, for the early cabins the foundation logs were directly placed on fieldstones (or sometimes also on sections of logs set vertically in the earth) at each of the four corners (Weslager C. A. 1969. *The Log Cabin in America. From Pioneer Days to the Present*, Rutgers University Press, Quinn &Boden Company, New Brunswick, New Jersey, p. 14).

v. see footnote xvi

Photo 6.   McPeake Cabin in Parkers Crossroads; detail – side chimney (photo T. Tomaszek, 2018).

Photo 7.   McPeake cabin during its reassembly at Parkers Crossroads; workers are using traditional woodworking technologies (photo Brandon McPeake, source: Genealogy Trails History Group. 2019. The McPeake Heritage. http://genealogytrails.com/tenn/henderson/mcpeakecrossroads3.html).

Photo 8. McPeake Cabin in Parkers Crossroads; detail – traditional way of embedding the
building (photo T. Tomaszek, 2018).

Photo 9. McPeake Cabin in Parkers Crossroads; detail – Portland Cement used to fill the
spaces between the logs (photo T. Tomaszek,
2018).

Due to present day legal requirements, and to make access easier for the disabled, a wheelchair ramp and new stairs were introduced to the reconstructed building, making it easier for visitors to access the interior of the cabin (Photo 10, 11). To reveal the full "original" context of the place, planners also designed the "farm's" surroundings. Therefore, they located the Corn Crib, a similarly relocated historical structure, near the house. Although it did not originally belong to the McPeake farm, the crib comes from yet another place near the former Parker`s House site. (Photo 12). This structure was also modified according to the concept of creating a historical farm and living space, as well as catering to tourist and visitor needs. The most important change was the extension of the roof on both sides (as in the case of the cabin) over an added seating area, creating a picnic place for visitors. Despite the changes, the basic form and the way of embedding of the Corn Crib structure was maintained in its original condition. (Photo 13).

Photo 10. McPeake Cabin in Parkers Cross-roads; detail – wheelchair ramp (photo T. Tomaszek, 2018).

Photo 11. McPeake Cabin in Parkers Cross-roads; detail – new stairs (photo T. Tomaszek, 2018).

Photo 12. Corn crib located near McPeake Cabin (photo T. Tomaszek, 2018).

Photo 13. Corn crib located near McPeake Cabin (photo T. Tomaszek, 2018).

Because of this interpretive direction, a type of "historical substance" was re-introduced to the Battlefield as well as a certain cohesive historical appearance. Something of the historical context was returned to the site. Specifically, the reconstruction of both the building and its surroundings was made to arouse feelings of "authenticity" in the viewer. Despite the availability of information about the reconstruction is included on the interpretive panels and in the available leaflets, the tourist is intentionally introduced into the re-created historical context.

3 CABIN AT MERIWETHER LEWIS MONUMENT (NATCHEZ TRACE PARKWAY)

The log Cabin at the Meriwether Lewis Monument (called sometimes The Meriwether Lewis Information Cabin or Grinder`s Cabin Museum) is located at milepost 385.9 on the Natchez Trace Parkway in Tennessee[vi]. It was constructed in the style of the early 1800`s thus it is a replica of the typical pioneer cabin from that time [National Park Service…, 2018, Natchez Trace….], (Photo 14).

Photo 14.   Cabin at the Meriwether Lewis Monument (photo T. Tomaszek, 2018).

The Meriwether Lewis National Monument was created to mark and commemorate the place where American explorer, Meriwether Lewis (of the Lewis and Clark expedition), died on October 11, 1809 as he was traveling along the road which later became part of the Natchez Trace Parkway [The Living New Deal, 2018]. This historical spot was already preserved around the mid-nineteenth century. A Monument was first erected in 1848 by the state of Tennessee over Lewis' grave site. Then, in 1925, the Meriwether Lewis National Monument was established by President Calvin Coolidge. Later, after the Natchez Trace Parkway was created in 1938, it became a key component of the Parkway [Meriwether Lewis, Undated].

Meriwether Lewis was born on August 18, 1774 to a plantation family in Charlottesville, Virginia [Wikipedia…, 2018, Meriwether Lewis…]. As a young man he began his career in the state police, then he continued in the army where he reached the rank of captain in 1801. At about this time Lewis received a proposal from President Thomas Jefferson for a post as his personal secretary. In all likelihood, President Jefferson was already then intending to appoint Lewis as the commander of the planned transcontinental expedition [Ambrose, 1996]. In order to properly prepare for this function in the years 1801-1803 Lewis studied science (inter alia at the University of Pennsylvania) [New Perspectives on The West, 2001]. The expedition finally began in 1803. For the co-commander of the expedition Lewis chose William Clark, with whom he served in the state police in 1795. The expedition was successful, and Clark and Lewis returned in 1806 having reached the Pacific coast in 1805. After resigning from military service, Lewis became Governor of the Louisiana Territory in 1808. In September 1809 he set out on a trip to Washington to respond to complaints addressed to him [Waldman, Wexler, 2004]. However, he did not reach his destination, as he died in a violent and mysterious way in the

vi. The Natchez Trace Parkway is a 444-mile recreational road and scenic drive through three states – Mississippi, Alabama and Tennessee. It roughly follows the "Old Natchez Trace", which is a historic travel corridor used by American Indians, "Kaintucks", as well as the European settlers, slave traders and soldiers (https://www.nps.gov/natr/index.htm). Meriwether Lewis National Monument is located at 191 Meriwether Lewis Park Rd., Hohenwald, TN 38462.

Grinder`s Stand, the inn run by the Griner family, which was located at the site of the current Monument in Hohenwald (70 miles southwest of Nashville, Tennessee). Whether he committed suicide, as Jefferson thought, or was murdered, as the family thought, is unknown [Guice, 2007]. Although he died young, Lewis played an integral role in shaping the American West [Meriwether Lewis, Undated].

Historians believe that the tavern owned by the Griner family (Robert Evans Griner and Priscilla Knight Griner) was established in 1807 and originally known as "Indian Line Stand". After the proprietors, the establishment was properly known "Griner`s Stand", but local people mistakenly, but permanently, began calling it "Grinder`s Stand" [Davis, 1995, p.32–33]. The proper use of the stand was to provide food and lodging to travelers passing through the Natchez Trace.

Most likely the original stand consisted of two rough log cabins that were adjoined at right angles with an additional space between them (so, it was most likely "dogtrot" type of building)[vii]. Both rooms had doors which were facing the Natchez Trace and one room also had a door facing the detached kitchen behind the stand. A barn and stable were also located on the property, although no one has determined their original locations [Davis, 1995, pp.32–33].

The log cabin in question was built in the 1930s[viii] by the Civilian Conservation Corps and is located less than 20 feet southwest of the original Grinder's Stand site [Wikipedia..., 2018, Grinder's....]. While the objective was to duplicate the design of original building, no reliable description of it could be found[ix] [The Living New Deal, 2018], therefore the cabin does not authentically match the design of the original building [Meriwether Lewis, Undated]. It was erected on the double-pen (or two rooms) floor plan and in the type of building known as "dogtrot"[x] [Jordan, 1985]. However instead of locating two chimneys on the sides of the building (as it was usually done in 1800`s), the cabin has one chimney in the middle of it, and on the central axis.

The wooden log Cabin at the Meriwether Lewis Monument was reconstructed in order to recreate a certain historical context and give a unique authenticity to the place of Lewis' death. Both the reconstruction of the building repeating the stylistic solutions characteristic for objects of this type from the early 1800s, as well as the layout and maintenance of the immediate surroundings (including the park surrounding to the Monument) is a far-reaching interpretation. As already mentioned, due to the lack of evidence about the original appearance of the place, the applied spatial and formal solutions created a secondary historical narrative, giving to this important place a dimension with a special emotional intensity.

The Cabin is located in the middle of a wooded area resembling the virgin terrain of Natchez Trace from the time of Lewis`s journey. The building is surrounded by old trees, and its setting fits perfectly with the landscape and the character of the place (Photo 15). The whole is seamed by elements of the fence, suggesting its authentic existence (Photo 16). And although one may doubt the validity of reconstructing a building to such shape but no other one, the most glaring interpretation issue seems to be the method of reconstruction and maintenance of the building. The Cabin imitates a historical object, although this imitation is not supported by the use of traditional techniques or materials (Photo 17, 18). This is evident, for example, in the method of processing of the surface of the logs. Particularly inept imitation of the surface traditionally

---

vii. See footnote i

viii. According to the National Park Service information it was built in 1935 (https://www.nps.gov/natr/ learn/historyculture/exploring-the-meriwether-lewis-site.htm)

ix. No drawings of Grinder's Stand are known to exist (https://www.nps.gov/natr/learn/historyculture/ exploring-the-meriwether-lewis-site.htm)

x. See footnote i

Photo 15. Cabin at the Meriwether Lewis Monument (photo T. Tomaszek, 2018).

Photo 16. Cabin at the Meriwether Lewis Monument (photo T. Tomaszek, 2018).

Photo 17. Cabin at the Meriwether Lewis Monument (photo T. Tomaszek, 2018).

Photo 18. Cabin at the Meriwether Lewis Monument (photo T. Tomaszek, 2018).

obtained in the treatment of wood by hand tools in the early 1800s can be noted on the elements exchanged during subsequent "repairs" of the building (Photo 19). They have specific vertical incisions suggesting traces of a machining tool. Slightly less visible, albeit made in a similar way, are identical imitations of machining tool on original elements dating from the reconstruction period (1930s) (Photo 20, 21). In addition, some elements "pretend" to be the exchanged elements, with the aim of creating the impression of subsequent repairs to the building. All this causes a certain chaos and lack of logic of the natural aging process of the wooden structure (Photo 22, 23).

Embedding of the building in the corners is also done in a dubious way. Support pillars are made up of masonry "pillars" (stones connected by contemporary mortar) instead of loosely laid fieldstones the traditional method[xi] [Weslager, 1969, p. 15], (Photo 24, 25). At the corner joints, the use of wood inserts to help shape the corner's dovetail joint log ends is an exceptionally inappropriate solution. They were made in all likelihood because the method of combining the elements during the reconstruction of the cabin was simplified and did not repeat typical, traditional carpentry solutions (Photo 26, 27).

Among many other historical alterations, modern technology and materials were used in the construction of the "dogtrot" between the rooms as well as the odd, centered chimney, (Photo 28), and again, Portland cement was used to fill the spaces between the logs forming

---

xi. See footnote iv

Photo 19.   Cabin at the Meriwether Lewis Monument; detail – vertical incisions imitating traditional treatment of wood by hand tools (photo T. Tomaszek, 2018).

Photo 20.   Cabin at the Meriwether Lewis Monument; detail – vertical incisions imitating traditional treatment of wood by hand tools (photo T. Tomaszek, 2018).

Photo 21.   Cabin at the Meriwether Lewis Monument; detail – vertical incisions imitating traditional treatment of wood by hand tools (photo T. Tomaszek, 2018).

Photo 22. Cabin at the Meriwether Lewis Monument; – "fake" repairs of the walls (photo T. Tomaszek, 2018).

the walls (which is a non-traditional solution[xii]) (Photo 27). The use of modern materials is noticeable all over the building, such as the application of metal "insulation" inserted at the building's entrance (Photo 28).

---

xii. See footnote xvi

Photo 23. Cabin at the Meriwether Lewis Monument; – detail – "fake" repairs of the walls (photo T. Tomaszek, 2018).

Photo 24. Cabin at the Meriwether Lewis Monument; – detail – not traditional embedding of the building in the corners (photo T. Tomaszek, 2018).

Photo 25. Cabin at the Meriwether Lewis Monument; – detail – not traditional embedding of the building in the corners; and "formation" of the shape of the endings of individual logs at the corners joints with the help of additional wood inserts (photo T. Tomaszek, 2018).

Photo 26. Cabin at the Meriwether Lewis Monument; – detail – "formation" of the shape of the endings of individual logs at the corners joints with the help of additional wood inserts (photo T. Tomaszek, 2018).

Photo 27. Cabin at the Meriwether Lewis Monument; – detail – "formation" of the shape of the endings of individual logs at the corners joints with the help of additional wood inserts; and fillings between the logs made from Portland cement (photo T. Tomaszek, 2018).

Photo 28. Cabin at the Meriwether Lewis Monument; – improper technologies used for "dogtrot"; secondary metal "insulation" inserted at the entrance to the building; and wheelchair ramp (photo T. Tomaszek, 2018).

Because the cabin is currently used as a small museum and information point, it must, by law, be accessible to disabled people. The method of making the ramp allowing easy and safe access for wheel chaired visitors may, however, be seen as undermining the authenticity of historical solutions (Photo 28). Of course, it remains an open question, what design and technology solution for such a ramp would minimize the falsification of the historical character of the building.

4   THE WILLIAM MANSE GEORGE CABIN

The William Manse George Cabin is located on the grounds of Shiloh National Military Park in Tennessee (Photo 29). It is simple one story building, erected on the single-pen floor plan as a variation of "English" and rectangular type [Jordan, 1985][xiii].

The Shiloh National Military Park was established on December 27, 1894 and then listed on the National Register of Historic Places on October 15, 1966 [Wikipedia..., 2019, Shiloh National....]. It preserves the scene of the first great American Civil War battle in the West – Shiloh and Corinth battlefields. In this 2-day battle, which took place on April 6 and 7, 1862, both Union and Confederate Armies suffered heavy casualties, as many as 24,000 people were killed, wounded or reported missing – a number equal to more than one-fifth of the combined two Armies engaged in the battle [Dillahunty, 1955, p. 1]. Because of the failure of Confederates to defeat the Federal Armies at Shiloh they were forced to return to Corinth, Mississippi and thus to relinquish all hold upon West Tennessee. Therefore the battle can be seen as a decisive victory for the Federal army when they advanced on and seized control of the Confederate railway system at Corinth [Cunningham, 2007, p. 32].

Photo 29.   The William Manse George Cabin in Shiloh National Military Park (photo T. Tomaszek, 2018).

The William Manse George Cabin is the only surviving structure of the more than seventy such buildings that resided on the battlefield in April 1862 [State of the Park

---

xiii. Single-pen (or one room) floor plan is the simplest shape of a wooden cabin. The type of it, which is on the square plan with centrally placed door openings at the front and rear as well as a chimney on the side wall outside is called "English" type, whereas one on a rectangular plan with door openings at the front and back moved to the side and a chimney on the side wall outside is called "Rectangular" type (Jordan T. G. 1985. *American Log Buildings. An Old World Heritage*, The University of North Carolina Press, Chapell Hill, p. 24)

Report..., 2016, p 26], [Dillahunty, 1955, p. 32, Paragraph 13]. According to the historical sources, this cabin formerly stood in Perry Field, near the main entrance to Shiloh Battlefield, approximately a mile north of its present location on the northern edge of Sarah Bell Field [State of the Park Report..., 2016, p 26]. During the first day of the battle, the last Confederate line was established just in the immediate front of the Cabin, whereas Federal army was located on its left. According to Dillahunty, "the battle-scarred logs reveal that it stood in the midst of heavy fighting" [Dillahunty, 1955, p. 32, Paragraph 13].

The building was relocated in the immediate aftermath of the battle on April 1862 to replace a cabin burned during the engagement [Dillahunty, 1955, p. 32, Paragraph 13]. Since 1895, when the building had a collapsed roof and crumbling walls, it has undergone several restorations. The last one to date was done in 2004. The work included, among other things, installation of a new roof and new mud and stick chimney [State of the Park Report..., 2016, p 26] (Photo 30, 31).

Photo 30.   The William Manse George Cabin after restoration from 2004 (author Mettendorf E., 2007, source: https://commons.wikimedia.org/wiki/File:W._Manse_George_Cabin,_Shiloh_National_Military_Park.JPG).

Photo 31.   The William Manse George Cabin after restoration from 2004 (author Bill Bechmann, 2012, source: http://www.civilwaralbum.com/shiloh/tour13_near.htm).

According to the information available (archival photo documentation), before 2004 the William Manse George Cabin did not have a side chimney, while from the front it was enriched with a covered porch (Photo 34, 35). Even though there is no comprehensive information about the appearance of the building which was originally located in this place and then destroyed during the battle, the reconstruction of this type of chimney seems to be consistent with the type used in the period of erecting the cabin.

Referring to the architectural historians, this kind of mud-and-stick chimney was typical for early wooden structures in Tennessee [Weslager, 1969, p. 15], [Rehder, 2012, p.23] (Photo 32, 33). However it is almost impossible to find surviving authentic chimney, because they either burned up during the early occupation of the building or were later replaced with limestone or brick chimneys.

Photo 32. Original mud-and-stick chimney at a saddlebag house in Hawkins County, Tennessee (after: Rehder J. B. 2012. *Tennessee log buildings: a folk tradition.* The University of Tennessee Press, Knoxville, p.23).

Photo 33. Stick-and-mud chimney typical for many early pioneer cabins; Replica of Fort Nashborough build around 1779, Nashville, Tennessee; (after: Weslager C. A. 1969. *The Log Cabin in America. From Pioneer Days to the Present.* Rutgers University Press, Quinn &Boden Company, New Brunswick, New Jersey, p. 8).

Although it is closed to daily entry by park visitors, the Cabin is a prime interpretive resource for the park. Therefore the park staff continue to use it when organizing interpretive programs on civilian life, particularly in regards to subsistence farming and impacts the war had on the surrounding rural community [State of the Park Report..., 2016, p 26].

The cabin is located in a picturesque place on the edge of an open space in the historical battlefield. In its background there is an old forest with, among other species, magnificent cedars. All these landscape elements taken together bring the observer into a context full of reflection and suggests that he has entered a place fixed in time (Photo 36, 37). The building – although originally located in a different part of the battlefield – filled the gap left by the previous building, thus it seems fully belonging to this place. Therefore its transfer takes on a different meaning due to the fact that it was a natural change in the local landscape caused by the events of the war.

Photo 34. The William Manse George Cabin before restoration in 2004 (author Tom Johnston, source: http://www.civilwaralbum.com/shiloh/tour13_near.htm).

Photo 35. The William Manse George Cabin before restoration in 2004 (author Carol Highsmith, source: http://www.loc.gov/pictures/item/2011630001/).

Photo 36. The William Manse George Cabin in Shiloh National Military Park (photo T. Tomaszek, 2018).

Photo 37. The William Manse George Cabin in Shiloh National Military Park (photo T. Tomaszek, 2018).

Despite many restorations (and resulting from it simultaneous changes in its shape) the state of falsification of the historical substance of the cabin seems almost imperceptible. The current condition of the building emphasizes its historic character, and this impression is enhanced by its natural patina. This process of ageing is noticeable for the average contemporary observer, for whom the original use of the peg marks on the wall is no longer obvious, but the existence of which is a witness to the passing of time (Photo 38).

Photo 38.   The William Manse George Cabin in Shiloh National Military Park; detail – traces of pegs on the walls (photo T. Tomaszek, 2018).

For the average visitor in direct confrontation with the cabin the conservation interventions are not obvious, and the previously executed restorations organically bond with the whole. Despite the fact that since the last restoration the reconstructed chimney was partially destroyed and is now preserved fragmentarily (Photo 39, 40), it does not give the impression of neglect. On the contrary, it evokes the state of a certain "authenticity" of the historical architectural structure ageing over the time.

The conservation treatments accomplished in the past have also not led in general to falsification of the original technologies. The building is embedded in an original way (that is, according to how it was practiced in times when it was built), on loosely arranged fieldstones placed in the joints of the corners[xiv] [Weslager, 1969, p. 15], (Photo 41). Similarly the stairs are made, allowing access to the building on both sides (Photo 42, 43). The roof covering is made of wooden shingles[xv] [Weslager, 1969, p. 15], (Photo 44), and the spaces

---

xiv. see footnote iv

xv. According to Weslager, the preferred roof covering for the early cabins were wooden shingles or "shakes" held down by "weight timbers". However their manufacture required adequate skills and riving tools (which many builders did not possess), and so they often covered roofs instead with pieces

Photo 39.   The William Manse George Cabin in Shiloh National Military Park; detail – partially destroyed chimney (photo T. Tomaszek, 2018).

Photo 40.   The William Manse George Cabin in Shiloh National Military Park; detail – partially destroyed chimney; still well preserved secondly introduced wooden elements from the time of reconstruction of chimney (photo T. Tomaszek, 2018).

Photo 41.   The William Manse George Cabin in Shiloh National Military Park; detail – traditional methods of embedding of the building in the corners (photo T. Tomaszek, 2018).

between the logs forming the walls are filled in a traditional way, with clay mixed with straws[xvi] [Weslager, 1969, p. 15], (Photo 45).

The completed conservation and restoration in the fill layer between the logs (which did not distort the look of the natural aging of the building) (Photo 46, 47), the new wooden elements (chimney) (Photo 40) as well as the exchanged wooden elements haven't disturbed the historical character of the cabin. Despite the natural patina (which is caused by natural factors like the action of variable humidity, UV radiation or wind) noticeable on the surface of wood there are still visible traces of manual woodworking (Photo 48). All this in whole

---

of bark or thatch (Weslager C. A. 1969. *The Log Cabin in America. From Pioneer Days to the Present*, Rutgers University Press, Quinn &Boden Company, New Brunswick, New Jersey, p. 15).

xvi. According to Weslager, the spaces between the logs forming the walls were traditionally filled with smaller pieces of wood or stones which were wedged tightly, and then caulked with moss, or wet clay, often tempered with animal hair or straw, in the process known as "daubing", or "chinking" (Weslager C. A. 1969. *The Log Cabin in America. From Pioneer Days to the Present*, Rutgers University Press, Quinn &Boden Company, New Brunswick, New Jersey, p. 15)

Photo 42. The William Manse George Cabin in Shiloh National Military Park; detail – stairs allowing access to the building (photo T. Tomaszek, 2018).

Photo 43. The William Manse George Cabin in Shiloh National Military Park; detail – stairs allowing access to the building (photo T. Tomaszek, 2018).

Photo 44. The William Manse George Cabin in Shiloh National Military Park; detail – roof covering made of wooden shingles (photo T. Tomaszek, 2018).

Photo 45. The William Manse George Cabin in Shiloh National Military Park; detail – traditional filling between the logs (photo T. Tomaszek, 2018).

Photo 46. The William Manse George Cabin in Shiloh National Military Park; detail – conserved and restored fill layer between the logs (photo T. Tomaszek, 2018).

Photo 47. The William Manse George Cabin in Shiloh National Military Park; detail – conserved and restored fill layer between the logs (photo T. Tomaszek, 2018).

Photo 48.   The William Manse George Cabin in Shiloh National Military Park; detail – visible traces of manual woodworking on the surface of logs (photo T. Tomaszek, 2018).

strengthens the feeling of the cabin's "authenticity," both its location and state of preservation.

5   THE FINAL REMARKS

The discussion above on the examples of the methods of interpretation and preservation of the historical wooden buildings directly related to the places of extremely important meaning for national identity are a contribution to the critical evaluation of both the transfer to a new location and the reconstruction of the architectural object, and thus the issue of its authenticity and the possibility of transmitting an unadulterated message on belonging to the place of a particular emotional charge.

The cases of McPeake Cabin in Parkers Crossroads and The William Manse George Cabin in Shiloh National Military Park illustrate the situation when the reinstatement of lost historical buildings is done by replacing them with other historical buildings from the same period and erected in similar style (shape, form, meaning, usage) as the original buildings. In the first case we deal with the issue of "using" a similar building, in the second case, it is an object belonging to the interpreted area, although originally from a different location. As discussed, the McPeake Cabin has undergone many transformations that not only distorted its original shape and meaning, but also transformed it into an artificially created building designed to imitate an "authentic" context. In view of the above, it is quite doubtful whether this intention has been properly implemented and the assumed effect achieved. At the same time, the question arises regarding maintenance of the authenticity of this historical structure. It seems to be indisputable that there has been both a loss of the authenticity of the place (in this case the cabin has been moved into another semantic space), as well as the authenticity of form (shape), materials or technological solutions. As a result, on the basis of a historic structure, it was created from a secondary building modified to meet the needs of both historical narrative and contemporary visitors.

The situation is different with the case of The William Manse George Cabin. Although the building was moved to a new site, it remained within the scope of its original semantic area. Thus, it seems to defend itself against criticism of improper translation of the senses (meanings) and falsification, and therefore to respond to the challenge of "authenticity" treated as belonging to the original context of the place and not just a specific point in space. Of course, the fact that formal transformations have been executed during subsequent restorations is debatable, but it seems that one can ascribe them into the natural sequence of existence of a building in this particular place, in the former battlefield.

The case of Cabin at Meriwether Lewis Monument located in Natchez Trace Parkway is undeniably the least convincing. This almost inept reconstruction of a certain architectural element of the original historical landscape, contradicting all principles of maintaining authenticity, seems to really only imitate the original meaning and cultural values of this place. Inadequate and adulterated spatial, material and technological solutions introduce the viewer into a confusion, and the artificially evoked emotional aspect associated with Lewis's death has nothing to do with an improper architectural structure.

6 SUMMARY

Presented and discussed above, the examples of reinstatement of lost historical buildings as part of the preservation and interpretation plan of a historical place of great importance for national identity show in a broad sense the essence of the issue of maintaining the authenticity of original context as well as authenticity of reinstituted and restored heritage structures. The principle of replacing a non-existing building with a similar structure that is practiced in a wide range raises a question about validation in general of the transfer of a building from its original place to new one. As this study has shown, this solution can be treated as compatible with the horizon of maintaining the authenticity of the object in question in the case of maintaining the original context of meaning (i.e. as in the case of The William Manse George Cabin in Shiloh National Military Park), that is only when "displacement" is executed in the area of identical historical, geographical and semantic meaning. In the situation of deprivation of the transferred building from this context, it comes to an inevitable falsification and the same to the creation of a newly interpreted historical tale.

Reconstruction from scratch, which is a reinstatement of both the architectural body itself and historical narrative, requires an adequate understanding of the correctness of the implemented solutions, because without their proper identification it leads to the inevitable modification of the original context of the historical message.

Considering the above remarks, it should finally be concluded that the wooden log cabin constituting a specific architectural archetype in the Tennessee area is also a direct witness of the history of this area. Thus, the appropriate preservation of the above heritage structures in accordance with the horizon of authenticity, especially when they are in themselves a form of interpretation of places of special historical importance, should be a priority task.

ACKNOWLEDGMENTS

The above research was carried out during the years of 2017/2018 under a Research Fellowship at the Center for Historic Preservation, Middle Tennessee State University, Murfreesboro, Tennessee, USA. It was also possible thanks to a research grant from the Kosciuszko Foundation.

Special thanks to Professor Carroll Van West, PhD; Tennessee State Historian, Director of MTSU Center for Historic Preservation, and Stacey Graham, Ph.D.; Research Professor at MTSU Center for Historic Preservation for their help and commitment in carrying out this research.

REFERENCES

Ambrose, S. E. 1996. *Undaunted Courage: Meriwether Lewis, Thomas Jefferson, and the Opening of the American West.* New York: Simon & Schuster. Print.

Bearss E., McDaniel S., Weaver J., and Roth D. (ed.).2003. Charge them both ways! The Battle of Parker`s Crossroads. Blue&Gray Magazine, Vol. XX, Issue 6 (Special Supplement to the Fall 2003 issue). http://www.parkerscrossroads.com/blue&gray/fall03weblo.pdf

Brown D. 1988. „What Really Happened to Meriwether Lewis?: The suicie explanation is not generalny believed down on the Natchez Trail". Columbia Magazine: „The Magazine Of Northwest History", Winter 1987–88: Vol. 1, No. 4, http://columbia.washingtonhistory.org/magazine/articles/1987/0487/0487--a1.aspx. (http://www.washingtonhistory.org/files/library/01-4_Brown.pdf) (accessed December 20, 2018).

Cunningham O. E. 2007. *Shiloh and the Western Campaign of 1862.* Ed. by Joiner G. and Smith T., New York: Savas Beatie.

Davis W. C. 1995. *A Way Through the Wilderness: The Natchez Trace and the Civilization of the Southern Frontier.* New York: Harper Collins.

Dillahunty A. 1955. Shiloh National Military Park, Tennessee. *National Park Service Historical Handbook Series* no. 10. Washington, D.C.

Eicher D. J. 2001. *The Longest Night: A Military History of the Civil War.* New York: Simon & Schuster.

Genealogy Trails History Group. 2019. The McPeake Heritage. http://genealogytrails.com/tenn/henderson/mcpeakecrossroads3.html (accessed January 14, 2019).

Guice J. D. W. (ed.). 2007. *By His Own Hand?: The Mysterious Death of Meriwether Lewis.* Norman: University of Oklahoma Press. Print.

Historical Marker Project. 2018. McPeake Cabin (HMDJ3). https://www.historicalmarkerproject.com/markers/HMDJ3_mcpeake-cabin_Wildersville-TN.html (accessed November 20, 2018).

Jordan T. G. 1985. *American Log Buildings. An Old World Heritage.* The University of North Carolina Press, Chapell Hill.

Meriwether Lewis. Undated. Natchez Trace Parkway. National Park Service. U.S. Department of the Interior. https://www.nps.gov/natr/planyourvisit/upload/M-Lewis-Site-Bulletin-8-30-13.pdf (accessed November 22, 2018).

National Park Service. U.S. Department of the Interior. 2018. Natchez Trace Parkway. https://www.nps.gov/natr/planyourvisit/tennessee.htm (accessed November 20, 2018).

National Park Service. U.S. Department of the Interior. 2018. Natchez Trace Parkway. Exploring the Meriwether Lewis Site. https://www.nps.gov/natr/learn/historyculture/exploring-the-meriwether-lewis-site.htm (accessed November 18, 2018).

New Perspectives on The West. 2001. Meriwether Lewis. https://www.pbs.org/weta/thewest/people/i_r/lewis.htm (accessed October 17, 2018).

Parker`s Crossroads. 2016. Battlefield Preservation. http://www.parkerscrossroads.org/battlefield-preservation/ (accessed November 26, 2018).

Rehder J. B. 2012. *Tennessee log buildings: a folk tradition.* The University of Tennessee Press. Knoxville.

Shedd, C. E. 1954. A history of Shiloh National Military Park., U.S. Department of the Interior, National Park Service, Report-2186760.

State of the Park Report for Shiloh National Military Park. 2016. *State of the Park Series* No. 25. National Park Service, Washington, DC.

The American Battlefield Protection Program (ABPP). 2015. Parker's Cross Roads. https://www.nps.gov/abpp/battles/tn011.htm (accessed September 18, 2018).

The American Battlefield Trust. 2019. The Battle of Parker's Cross Roads: Then & Now. https://www.battlefields.org/learn/articles/battle-parkers-cross-roads-then-now (accessed January 21, 2019).

The CivilWarTalk. 2010 – 2018. Nathan B. Forrest. https://civilwartalk.com/threads/forrest-at-parkers-crossroads-by-john-paul-strain.93296/(Accessed January 13, 2019).

The Living New Deal. 2018. Meriwether Lewis National Monument – Hohenwald TN. https://livingnewdeal.org/projects/meriwether-lewis-national-monument-hohenwald-tn/ (accessed December 13, 2018).

The Parker`s Crossroads Battlefield Association. 2000-2012. The Battle at Parker's Crossroads, Tennessee. http://www.parkerscrossroads.com/Battle_Information/history.htm (accessed January 12, 2019).

Waldman C., Wexler A. 2004. *Encyclopedia of Exploration*. T. I: The Explorers. New York: FactsOnFile.

Weslager C. A. 1969. *The Log Cabin in America. From Pioneer Days to the Present*, Rutgers University Press, Quinn & Boden Comp., New Brunswick, New Jersey.

Wikipedia. The Free Encyclopedia. 2018. Grinder`s Stand. https://en.wikipedia.org/wiki/Grinder%27s_Stand (Accessed September 2, 2018).

Wikipedia. The Free Encyclopedia. 2018. Meriwether Lewis. https://pl.wikipedia.org/wiki/Meriwether_Lewis (accessed November 12, 2018).

Wikipedia. The Free Encyclopedia. 2019. Battle of Parker`s Cross Roads. https://en.wikipedia.org/wiki/Battle_of_Parker%27s_Cross_Roads (accessed January 17, 2019).

Wikipedia. The Free Encyclopedia. 2019. Shiloh National Military Park. https://en.wikipedia.org/wiki/Shiloh_National_Military_Park (accessed January 29, 2019).

The preliminary recognition of the condition and authenticity of historical log structures remaining in Cane Ridge community (Antioch, Tennessee)

ABSTRACT: This paper presents and discusses in terms of the authenticity of heritage buildings the significant historical log structures still existing in the Cane Ridge community, a few miles southeast of Nashville proper, and within the suburban township of Antioch, Tennessee. As the traditional type of vernacular architecture of the first settlers in Middle Tennessee, these wooden structures are an integral component of the cultural landscape of this region, reflecting the identity of its inhabitants. Therefore, the proper protection and preservation of these buildings for future generations is of the utmost importance. To determine adequate conservation methodology, it is necessary to recognize and understand the authenticity of these structures. Such an analysis is essential for identifying processes that have modified their original formal, spatial and technological solutions.

1   INTRODUCTION

The aim of this project was to locate and conduct preliminary research in terms of the authenticity of the significant historical log structures still existing in the Cane Ridge community, a few miles southeast of Nashville proper, and within the suburban township of Antioch, Tennessee. These buildings, now disappearing at an accelerated pace from the cultural landscape of Middle Tennessee due to development pressure, were once the basic type of vernacular architecture of these lands.

The reconnaissance allowed the development of a general assessment of these cultural and structural features, and thus we obtain a clearer picture of the layered history and building tradition of the overall community. However, due to the complex nature of the issue, it should be remembered that the analyses only of individual buildings are in fact selective and at the same time "a-historical" and thus insufficient for recognizing the complex integrative nature of rural community life and its building tradition [McMurry, 2001, p. 34]. Thus, broader understanding of the issue requires further diagnoses of context and studies of cultural meaning, allowing for a fuller reading of the complex character of the problem, to which the present analyses constitute only a contribution.

The log structures presented in this paper are a kind of specific chronicle of changes that have occurred both in local building tradition and the way of life of their residents. The analysis of the context of the authenticity of these buildings remains an extremely difficult and multilayered task (as it requires the proper recognition and understanding of historical architectural and technological solutions while capturing the dynamics of later transformations at the formal and aesthetic level), however, its execution is essential for adequate preservation of the material culture and built environment of the region in question.

2   HISTORICAL LOG STRUCTURES IN THE CANE RIDGE COMMUNITY

The investigations made it possible to determine that a surprising number of historical log structures or their remains are still present in the Cane Ridge and directly adjacent areas, and unexpectedly many of them are early structures, dating back even to 1805.

According to the local lore, several buildings may even be older, including the log portions of the Gillespie-Culbertson house and the Daniel Gray Clark house, both purported (by their current owners) to date to the late 1700s, as well as log structures imbedded within or obscured by later board siding, such as the Burkitt and Whitsett houses. [Andrews, 2017,p. 4]

Several of them are still owned by descendants of early residents or original builders. What's also interesting, the surviving percentage of the historical fabric essential for this cultural landscape of Middle Tennessee is especially significant given the relatively small size of the community and the degree of development that has occurred in the area[i]. Unfortunately only one home in the immediate area, the Benajah Gray Log house, is currently listed on the National Register of Historic Places, receiving that designation in 1985. [Andrews, 2017, p. 4]

This paper describes the eight historical log structures essential for understanding the current situation of historical preservation and historical wooden buildings in this area of Middle Tennessee. There are as follows: Benajah Gray house and Slave Dwelling, Thomas Johnson House, B. Barnes house, Francis Waller house, Whitsett-Turner-Hill house, Gillespie-Culbertson house, Ledbetter Cabin, King Cabin.

3   BENAJAH GRAY HOUSE

The Benajah Gray house (Photo 1, 2), located at 6320 Burkitt Road, was constructed circa 1805. It is a double-pen with a central hallway, having a single-pen one-story structure on one side (sheathed in board siding) and a single-pen one-and-a-half story structure on the other side, all constructed at the same time of diamond-notched yellow poplar and cedar logs [Clements, 1987, vol. II, p. 103] [Andrews, 2017, p. 5] (Photo 3) According to the description in the National Register nomination, the simultaneous construction of two adjoining pens of different heights as well as the diamond notching are atypical for the region [National Register..., 1985] [Andrews, 2017, p. 5].

Photo 1.   Benajah Gray house (photo T. Tomaszek, 2017).

Photo 2.   Benajah Gray house (photo T. Tomaszek, 2017).

---

i. In addition to the architectural structures discussed below, it is worth mentioning additional important log structures in the community, like Wolf House, Smothers-Culbertson House, Buchanan Tavern and Burkitt House, which reflect the richness of wooden building tradition of this area.

Photo 3. Benajah Gray house – Diamond Notches type of corner joint (photo T. Tomaszek, 2017).

Photo 4. Benajah Gray house – the back of the building (frame kitchen currently connected with the log part by the later roofing) (photo T. Tomaszek, 2017).

At the back of the building there is a one-story shed, which is later addition and was originally a porch remodeled at some point into a kitchen. In the 1870s, a separate one-story frame building with a central chimney was also erected. It served as a larger detached kitchen and at the same time a dining space. According to information obtained from the current owner of the building and a Gray descendant, Vicki Jordan, it was also used for entertaining and dances[ii] [Andrews, 2017, p. 5] (Photo 4).

Among other components of the property it should be mentioned as follow: a well, made of limestone and cinder block, no longer in use and covered, in front of the frame kitchen building; two smaller log structures behind the house, described in the National Register nomination as a kitchen (Photo 7, 8) and a smokehouse (Photo 5, 6); a twentieth-century framed pole barn; a family cemetery; and a one-and-a-half-story, double-pen, central-chimney slave house located on the south side of Burkitt Road, now within Cane Ridge Park [National Register..., 1985]; [Clements, 1987, vol. II, p. 224], [Andrews, 2017, p. 5].

Photo 5. Benajah Gray house – smokehouse, one of the two smaller log structures behind the house (photo T. Tomaszek, 2017).

Photo 6. Benajah Gray house – smokehouse, one of the two smaller log structures behind the house (photo T. Tomaszek, 2017).

ii. The information are from the personal communication of Jenny Andrews with Vicki Jordan, which took place on the property in Antioch, TN, on September 20, 2016.

Photo 7. Benajah Gray house – kitchen, one of the two smaller log structures behind the house (photo T. Tomaszek, 2017).

Photo 8. Benajah Gray house – kitchen, one of the two smaller log structures behind the house (photo T. Tomaszek, 2017).

Photo 9. Benajah Gray house – interior (photo T. Tomaszek, 2017).

Photo 10. Benajah Gray house – interior (photo T. Tomaszek, 2017).

Photo 11. Benajah Gray house – interior (photo T. Tomaszek, 2017).

Photo 12. Benajah Gray house – interior (photo T. Tomaszek, 2017).

4   THOMAS JOHNSON HOUSE

The Thomas Johnson house, (Photo 13, 17) located at 4011 Twin Oaks Lane, was constructed in 1830 by Thomas Johnson, considered to be a "master carpenter" [Clements, 1987, vol. I, p. 136] (Photo 15). The house is described as a two-story single-pen of half-dovetail log construction, with a central-bay, two-story front porch (Photo 14, 16), and a later one-story log ell addition at the back [Clements, 1987, vol. II, p. 105] (Photo 18, 19, 21). According to the family lore, it seems that at least part of the addition appears to have been a detached kitchen, with a later frame connector joining it to the main house [Andrews, 2017, p. 6].

Photo 13. Thomas Johnson house (photo T. Tomaszek, 2017).

Photo 14. Thomas Johnson house – two-story front porch (photo T. Tomaszek, 2017).

Photo 15. Thomas Johnson house – badge with construction date (photo T. Tomaszek, 2017).

Photo 16. Thomas Johnson house (photo T. Tomaszek, 2017).

Photo 17. Thomas Johnson house – log construction part (left); later addition at the back (right) (photo T. Tomaszek, 2017).

Photo 18. Thomas Johnson house – later addition at the back (photo T. Tomaszek, 2017).

Photo 19. Thomas Johnson house – later one-story log ell addition at the back (photo T. Tomaszek, 2017).

Photo 20. Thomas Johnson house – log construction part, detail (photo T. Tomaszek, 2017).

Photo 21. Thomas Johnson house – later one-story log ell addition at the back; detail (photo T. Tomaszek, 2017).

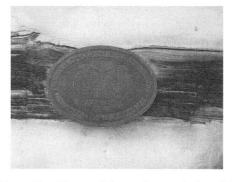

Photo 22. Thomas Johnson house – badge with Architectural Award from Metropolitan Historical Commission, Nashville, Davidson County (photo T. Tomaszek, 2017).

Photo 23. Thomas Johnson house – log construction part, detail (photo T. Tomaszek, 2017).

Photo 24. Thomas Johnson house – log construction part, detail (photo T. Tomaszek, 2017).

In the 1930s the house was already in disrepair. In 1937 it was purchased at auction by descendant Ben Paul, who subsequently sold it to his daughter Fannie Belle Taylor. Referring to the historical sources she restored the house, and – according to the current owner Eddie Paul – it was done properly, as the house was carefully dismantled, repaired, and reassembled [Clements, 1987, vol. I, p. 138], [Andrews, 2017, p. 6]. Today the Paul family still maintains the property, now only a little over an acre, though no one resides in the house full time [Andrews, 2017, p. 6].

5   B. BARNES HOUSE

B. Barnes house, or possibly R. Barnes house (Photo 25, 26, 29, 35), which is located at 6375 Pettus Road, seems to have been the home of the Barnes family, [Andrews, 2017, p. 9] and most likely built around 1844. This is evidenced by the date written on one of the wooden elements that was found during repairs made to the chinking after tornado damage in 1974. [Andrews, 2017, p. 10]. The name B. (or R.) Barnes appears next to the building's location on an important map of Davidson County, Tennessee from 1871, the Wilbur F. Foster map [Foster map, 1871]. The house is a one-an-a-half-story dogtrot with half-dovetail notched yellow poplar logs (Photo 27, 32, 33) and a chestnut log interior. [Andrews, 2017, p. 10] [Clements, 1987, vol. II, p. 223] (Photo 36, 37, 38). The original stone chimneys were decimated and the kitchen addition at the rear was damaged by the 1974 tornado [Andrews, 2017, p. 10]. (Photo 30, 34). It seems however that the original structural components are mostly intact, even though some necessary changes were made after tornado damage [Andrews, 2017, p. 10]

Photo 25.   B. Barnes house (photo T. Tomaszek, 2017).

Photo 26.   B. Barnes house – detail (photo T. Tomaszek, 2017).

Photo 27. B. Barnes house – detail – Half-Dovetail Notched type of corner joints (photo T. Tomaszek, 2017).

Photo 28. B. Barnes house – detail – filling between the logs after conservation (photo T. Tomaszek, 2017).

Photo 29.   B. Barnes house (photo T. Tomaszek, 2017).

Photo 30. B. Barnes house – detail (photo T. Tomaszek, 2017).

Photo 31. B. Barnes house – detail (photo T. Tomaszek, 2017).

Photo 32. B. Barnes house – detail – filling between the logs after conservation (photo T. Tomaszek, 2017).

Photo 33. B. Barnes house – detail – the traces of original, traditional hand-processing of the surface of wooden logs (photo T. Tomaszek, 2017).

Photo 34. B. Barnes house – detail (photo T. Tomaszek, 2017).

Photo 35. B. Barnes house (photo T. Tomaszek, 2017).

Photo 36. B. Barnes house – interior (photo T. Tomaszek, 2017).

Photo 37. B. Barnes house – interior (photo T. Tomaszek, 2017).

Photo 38. B. Barnes house – interior (photo T. Tomaszek, 2017).

6 FRANCIS WALLER HOUSE

The Francis Waller house (Photo 39, 40) is located at 6431 Pettus Road and most likely was built in the second part of nineteenth century. The width of the logs and use of yellow poplar does seem to indicate a date prior to 1880 [Rehder, 2012, p. 14]. In the written sources it is designated as the "Frank Walker" house [Clements, 1987, vol. II, p. 223], most likely based on the above-mentioned Foster map from 1871. [Andrews, 2017, p. 10]. However, according to Jenny Andrews, Clements misread the name on the map, which is actually "F. Wallers" and likely refers to Francis Waller whose name appears in nineteenth-century records for the area, such as the 1860 census (though the 1870 census lists him as "Francis Walker" and the 1850 census lists the surname as "Warler"). [Foster map, 1871] [Andrews, 2017, p. 11]. Since the nineteenth century, the Waller family has operated a funeral home in nearby Nolensville, but they were also cabinetmakers as well as makers and repairers of carriages. [Andrews, 2017, p. 11]. The house is one-and-a-half stories of half-dovetail notched yellow poplar logs (Photo 41), with a 1919 frame addition (Photo 42) [Clements, 1987, vol. II, p. 223].

Photo 39. Francis Waller house (photo T. Tomaszek, 2017).

Photo 40. Francis Waller house (photo T. Tomaszek, 2017).

Photo 41. Francis Waller house – detail – Half-Dovetail Notched type of corner joints (photo T. Tomaszek, 2017).

Photo 42. Francis Waller house – detail – frame addition from 1919 (photo T. Tomaszek, 2017).

Photo 43. Francis Waller house – interior (photo T. Tomaszek, 2017).

Photo 44. Francis Waller house – interior (photo T. Tomaszek, 2017).

Photo 45. Francis Waller house – interior (photo T. Tomaszek, 2017).

Photo 46. Francis Waller house – interior (photo T. Tomaszek, 2017).

Photo 47. Francis Waller house– interior (photo T. Tomaszek, 2017).

Photo 48. Francis Waller house – interior (photo T. Tomaszek, 2017).

Photo 49. Francis Waller house – interior (photo T. Tomaszek, 2017).

7 THE WHITSETT (TURNER-HILL) HOUSE

The Whitsett House (Photo 50, 51), located at 6947 Burkitt Road, was most likely built in the first half of nineteenth century. It is an early two-story double-pen log structure [Clements, 1987, vol. II, p. 223], that many years since has been sheathed in white board siding [Andrews, 2017, p. 11]. The home`s early owners (possibly the first owners) were William Whitsett, pastor of the nearby Concord Church, beginning in the 1840s, and his wife Malinda Turner Whitsett. [Andrews, 2017, p. 11]. This structure (building) is also associated with the Hill and Turner families (who were related to the Whitsett family), thus it is sometimes called the Whitsett-Turner-Hill house [Andrews, 2017, p. 11].

Photo 50. Whitsett (Turner-Hill) house (photo T. Tomaszek, 2017).

Photo 51. Whitsett (Turner-Hill) house (photo T. Tomaszek, 2017).

Photo 52. Whitsett (Turner-Hill) house (photo T. Tomaszek, 2017).

Photo 53. Whitsett (Turner-Hill) house – interior (photo T. Tomaszek, 2017).

Photo 54. Whitsett (Turner-Hill) house – interior (photo T. Tomaszek, 2017).

Photo 55. Whitsett (Turner-Hill) house – interior (photo T. Tomaszek, 2017).

A frame, single-room, cabin sized extension, possibly an early kitchen (and perhaps log underneath the siding) is attached to the back of the house (Photo 52). Behind the house are a log smokehouse with a front-facing projecting gable and a side lean-to gear shed with top-hinged wooden "windows" (Photo 56, 57, 58, 59, 60, 61), a small chicken coop, and an outhouse. Visible elsewhere on the property are various sheds and barns, and a silo [Andrews, 2017, p. 12] (Photo 62, 63).

Photo 56. Whitsett (Turner-Hill) house – log smokehouse with a front-facing projecting gable and a side lean-to gear shed (photo T. Tomaszek, 2017).

Photo 57. Whitsett (Turner-Hill) house – log smokehouse with a side lean-to gear shed – detail (photo T. Tomaszek, 2017).

Photo 58. Whitsett (Turner-Hill) house – log smokehouse with a side lean-to gear shed – detail (photo T. Tomaszek, 2017).

Photo 59. Whitsett (Turner-Hill) house – log smokehouse; detail – corner joints (photo T. Tomaszek, 2017).

Photo 60. Whitsett (Turner-Hill) house – log smokehouse with a side lean-to gear shed – detail (photo T. Tomaszek, 2017).

Photo 61. Whitsett (Turner-Hill) house – log smokehouse with a side lean-to gear shed with top-hinged wooden "windows" (photo T. Tomaszek, 2017).

Photo 62. Whitsett (Turner-Hill) house – barn located at the property (photo T. Tomaszek, 2017).

Photo 63. Whitsett (Turner-Hill) house – barn and silo located at the property (photo T. Tomaszek, 2017).

8   THE GILLESPIE-CULBERTSON HOUSE

The Gillespie-Culbertson house (Photo 64, 68) is located at 3402 Old Franklin Road. According to its current owner, Theola Holloway, the oldest part of it, the log structure, was constructed in the late 1700s as the original homestead[iii] [Andrews, 2017, p. 12] (Photo 65). It seems to be quite possible, as the size of the log structure and its upstairs loft (whose supports are visible on the exterior of the building), indicate more dwelling than a detached kitchen [Andrews, 2017, p. 12].

The main part of the house, and the most dominant, is a white, two-story frame I-house, with a central-bay two-story front porch with milled detailing that could be described as Italianate [Clements, 1987, vol. II, p. 224], though it might also be considered Folk Victorian [McAlester, 2015, p. 402] (Photo 66, 67). It was built by Andrew Gillespie (1849-1882) and later owned by the Culbertsons [Brown Johnson, 1973, p. 134].

---

iii. The information comes from the personal communication of Jenny Andrews with Theola Holloway, which took place on the property in Antioch, TN, on September 20, 2016.

Photo 64. Gillespie-Culbertson house (photo T. Tomaszek, 2017).

Photo 65. Gillespie-Culbertson house – the oldest part, log structure (photo T. Tomaszek, 2017).

Photo 66. Gillespie-Culbertson house – the main part, white, two-story frame I-house (photo T. Tomaszek, 2017).

Photo 67. Gillespie-Culbertson house – the main part, white, two-story frame I-house (photo T. Tomaszek, 2017).

Photo 68. Gillespie-Culbertson house (photo T. Tomaszek, 2017).

Photo 69. Gillespie-Culbertson house – interior (photo T. Tomaszek, 2017).

Photo 70.   Gillespie-Culbertson house – interior (photo T. Tomaszek, 2017).

Photo 71.   Gillespie-Culbertson house – interior (photo T. Tomaszek, 2017).

As mentioned above, the oldest part, the log structure, is at the rear of the later frame building, now joined to the main house by a frame connector. It is a one-and-a-half-story log structure with a stone chimney [Andrews, 2017, p. 12] (Photo 72, 73). The name of the first owner, and probably the builder of this log structure is unknown. On the Foster Map from 1871 a structure is indicated exactly where the Gillespie-Culbertson house is standing, however there is no name assigned to it and it is not clear if it applies only to the log part or the whole house[iv] [Foster map, 1871]. One explanation for not giving the name of the house on the map is that no one was living in the house in 1871, or an African American family was residing there at the time of the map survey [Andrews, 2017, p. 13]. Since, according to the Foster Map, it was located near two homes belonging to members of the Ham family in the 1870s, it is quite possible that the log structure was originally part of the Ham estate [Foster map, 1871]. The 1860 slave census lists five members of the Ham family owning a total of

Photo 72. Gillespie-Culbertson house – the oldest part, log structure (photo T. Tomaszek, 2017).

Photo 73. Gillespie-Culbertson house – the oldest part, log structure (photo T. Tomaszek, 2017).

---

iv. In all likelihood at the time of the 1871 map, Andrew Gillespie had not yet built the I-house section. He is listed in the 1870 census as living in another household nearby; he married in 1871 and had six children before his death in 1882. So it is likely that he built the I-house in the early to mid 1870s. According to the historical sources Andrew Gillespie purchased about 26 acres from Bradley Ham in 1872 [Andrews, 2017].

eleven slaves and six slave houses (which, compared to other slave owners in Cane Ridge, is a high number of houses per number of slaves), so perhaps the log portion of the Gillespie-Culbertson house was initially used as enslaved quarters [Andrews, 2017, p. 14].

On the property, behind the house, is located a sandstone dependency. According to Theola Holloway it is a smokehouse[v], though a stone smokehouse would be unusual for the area, and the examination of the interior revealed no hooks or scorch marks, which are typical indicators of smokehouse use. To the southeast of the main house and across the yard from the smokehouse there is also a stone well [Andrews, 2017, p. 12].

9 LEDBETTER CABIN

This small log building, which is one single-pen one-story log structure with saddle-notch construction, is located behind a mid-twentieth-century brick house on Cane Ridge Road (Photo 74, 75, 76, 77).

Photo 74. Ledbetter cabin, located behind a mid-twentieth-century brick house (photo T. Tomaszek, 2017).

Photo 75. Ledbetter cabin (photo T. Tomaszek, 2017).

It is unknown when it was erected, though it appears to be just east of the "W. Moore" (probably William Green Moore) house on the 1871 Foster Map[vi] [Foster map, 1871] [Andrews, 2017, p. 18], so therefore it could be probably dated to approximately the late-nineteenth century. It is also unknown if the structure was originally built as a dwelling, and if so whether it was intended for whites or blacks [Andrews, 2017, p. 19]. It is even more uncertain, as slave schedules do not show the Moores as slave owners [Andrews, 2017, p. 19]. However, some clues to the history of the building and its earlier forms and uses are found in the analysis of its elements as well as structural changes, especially in relation to windows and doors. The logs are not large, some are roughly hewn while others were left whole, and a large opening beside the chimney has been awkwardly filled with a crudely installed double window (Photo 78, 79). These components could indicate the structure was constructed in the late nineteenth century or early twentieth century, and even perhaps not originally intended as a dwelling.

---

v. The information comes from the personal communication of Jenny Andrews with Theola Holloway, which took place on the property in Antioch, TN, on September 20, 2016.

vi. The deeds indicate that the Ledbetters purchased property that had once belonged to William Green Moore [Andrews, 2017].

Photo 76.   Ledbetter cabin (photo T. Tomaszek, 2017).

Photo 77.   Ledbetter cabin (photo T. Tomaszek, 2017).

Photo 78.   Ledbetter cabin – detail – double window installed in large opening beside the chimney (photo T. Tomaszek, 2017).

Photo 79.   Ledbetter cabin – detail – elevation with chimney and the double window installed in large opening beside it (photo T. Tomaszek, 2017).

However, other conclusions can be drawn from the fact of the proximity from the building of an unusual, barrel-shaped, hand-cranked apparatus attached to a brick-and concrete base, used for drawing water from the well below [Andrews, 2017, p. 18] (Photo 80). It suggests a direct association with a dwelling and that the building could be in its original spot and not moved from somewhere else [Andrews, 2017, p. 18].

According to current owner Daniel Smith, the house was most recently occupied in the early twentieth century by two sisters, up until the brick ranch house was constructed at mid-century[vii]. This information is complemented by the neighbor Paul Wingler, who states that the property was owned by Sara Ledbetter Lynch who lived in the brick home, and whose widowed sister Mable Ledbetter Bell lived in the log house [Andrews, 2017, p. 18].

The interior of the log building does indicate use as a living space during the twentieth century, including linoleum flooring, plaster drywall, a dropped ceiling, no fireplace but a round opening for a stovepipe (venting into a chimney made of tool-marked sandstones

---

vii. The information comes from the personal communication of Jenny Andrews with Daniel Smith, which took place on the property in Antioch, TN, on December 30, 2016.

Photo 80. Ledbetter cabin – an unusual, barrel-shaped, hand-cranked apparatus attached to a brick-and concrete base, used for drawing water from the well below, located close to building, on its left side (photo T. Tomaszek, 2017).

at the base and cinderblocks for the shaft), and the remains of kitchen cabinets in the lean-to addition at the back of the building, which served as a kitchen (Photo 81, 82).

At some point in the latter half of the twentieth century a fire sparked outside the front of the building, which the homeowners rushed to extinguish; the scorch marks are still visible [Andrews, 2017, p. 19] (Photo 83, 84, 85, 86).

Photo 81. Ledbetter cabin – interior (photo T. Tomaszek, 2017).

Photo 82. Ledbetter cabin – detail – the lean-to addition at the back of the building, which served as a kitchen (photo T. Tomaszek, 2017).

Photo 83. Ledbetter cabin – the scorch marks noticeable on the surface of exterior walls (photo T. Tomaszek, 2017).

Photo 84. Ledbetter cabin – the scorch marks noticeable on the surface of exterior walls (photo T. Tomaszek, 2017).

Photo 85.   Ledbetter cabin – the scorch marks noticeable on the surface of exterior walls (photo T. Tomaszek, 2017).

Photo 86.   Ledbetter cabin – the scorch marks noticeable on the surface of exterior walls (photo T. Tomaszek, 2017).

10   KING CABIN

This small log structure is located at 6036 Culbertson Road, at the site and near the historical home currently owned by Judith King Vulcano (Photo 87, 88).

Photo 87.   King cabin (photo T. Tomaszek, 2017).

Photo 88.   King cabin (photo T. Tomaszek, 2017).

The home is constructed from a combination of log and stone. The log part of the home was built in 1938 (this date is provided by the property assessor`s website), and according to Teresa King Couch (the sister of Judith), it was erected from the logs acquired by Jesse King (grandfather of Teresa and Judith) in about 1936 from an old warehouse being torn down in downtown Nashville [Andrews, 2017, p. 20] (Photo 89). The original logs were quite large, and Mr. King had them cut lengthwise to a smaller size, using them to build a small house. [Andrews, 2017, p. 20]. The other half of the house is of crab orchard stone and built in 1951, according to a photograph owned by Judith, which is dated and shows the crab orchard-stone structure nearing completion [Andrews, 2017, p. 20].

Photo 89.   King's family house (owned currently by Judith King Vulcano), located at the same property with King cabin – detail – the wooden log part (photo T. Tomaszek, 2017).

Referring to the family lore, a log cabin now located near the home had originally belonged to one of their ancestors, John W. King, who was great-great-grandfather of Teresa and Judith King, and who built it about 1835 (according the notations accompanying family photos of the cabin). Originally it was located west of Nolensville, on the south side of Clovercroft Road, in nearby Williamson County. In the 1960s the cabin was about to be torn down, so Jesse King (Teresa and Judith's father) together with his friend Lester Scales decided to move it to the current location. The two men marked the logs, took the building apart, and reconstructed it beside the Jesse King house in 1965 [Andrews, 2017, p. 21].

The cabin is one-and-a-half stories, single pen (Photo 90, 91), with half-dovetail notched logs (Photo 94), one end chimney of limestone (Photo 92, 93), and an interior staircase to the upstairs loft (Photo 98, 99). Apparently, the limestone rocks for the chimney (and maybe the foundation) accompanied the cabin and tool markings are evident. Inside is a salting trough made from a huge log, which potentially came from the same location as the cabin [Andrews, 2017, p. 21] (Photo 97).

Photo 90.   King cabin (photo T. Tomaszek, 2017).

Photo 91.   King cabin (photo T. Tomaszek, 2017).

Photo 92. King cabin (photo T. Tomaszek, 2017).

Photo 93. King cabin – detail (photo T. Tomaszek, 2017).

Photo 94. King cabin – detail – Half-Dovetail Notched type of corner joints (photo T. Tomaszek, 2017).

Photo 95. King cabin – detail – filling between the logs (photo T. Tomaszek, 2017).

Photo 96. King cabin – detail – filling between the logs (photo T. Tomaszek, 2017).

Photo 97. King cabin – interior – detail – salting trough made from a huge log (photo T. Tomaszek, 2017)

Photo 98. King cabin – interior (photo T. Tomaszek, 2017).

Photo 99. King cabin – detail – an interior staircase to the upstairs loft (photo T. Tomaszek, 2017).

11   HISTORICAL MODIFICATIONS OF THE BUILDING STRUCTURE AND THE HORIZON OF AUTHENTICITY OF THE HISTORICAL BUILDING

The existing historical log structures are an important element of the Cane Ridge community`s cultural landscape. It is, therefore, essential to protect them in the way that guarantees that their authentic historical, formal or material dimension will be maintained. Thus, it is necessary not only to recognize the original formal or technological solutions, but also the changes that shaped the architectural structure of individual buildings throughout their historical existence.

Most of the log structures still existing at Cane Ridge have undergone significant modifications since their construction. This applies mainly to changes made both in the architectural shape and the function of a given building. Often, the original structure has been enclosed by later additions, which is of course related to the need to increase the usable area of the house and to adapt it to the requirements of a given historical time, changing over the subsequent decades. This is the case in most of the existing residential log structures in Cane Ridge. An interesting example of a gradual and consistent expansion of the architectural shape/space based on the original, primary log structure may be, for example, the Benajah Gray house (Photo 100), Thomas Johnson house (Photo 101), B. Barnes house (Photo 102), Francis Waller house or the Whitsett (Turner-Hill) house (Photo 103).

Photo 100. Benajah Gray house (photo T. Tomaszek, 2017).

Photo 101. Thomas Johnson house (photo T. Tomaszek, 2017).

Photo 102. B. Barnes house (photo T. Tomaszek, 2017).

Photo 103. Whitsett (Turner-Hill) house (photo T. Tomaszek, 2017).

In the consecutive transformations carried out over time, occupants used the architectural solutions typical for the era/period in which the changes were executed, ipso facto, as a consequence, no direct reference was made to formal or technological solutions used in the original/primary part of the structure, i.e. the part that is the log structure. The architectural solid/form created in this way is nowadays often an eclectic conglomeration of various architectural and stylistic solutions, additionally usually characterized by the diversity of materials and technologies used (Photo 104, 105).

Photo 104. Gillespie-Culbertson house – noticeable the eclectic conglomeration of various architectural and stylistic solutions, as well as the diversity of materials and technologies used (photo T. Tomaszek, 2017).

Photo 105. Gillespie-Culbertson house – noticeable the eclectic conglomeration of various architectural and stylistic solutions, as well as the diversity of materials and technologies used (photo T. Tomaszek, 2017).

Because of the changing historical consciousness and accelerated recent processes of urbanization and the extensive development of rural communities in most parts of the US, historical vernacular structures that survive today take on special significance. Once a core of the local building tradition and the local cultural landscape, log structures today are a testimony to the past and radical changes related to the processes of progressive globalization.

As the objects constituting the identity of the local community, re-recognized and appreciated by this community, these structures, thanks to their intended exposure, become an extremely important element of the manifestation of local cultural and historical traditions (Photo 106). As a form of expressing a particular emotional message, their manner of preservation and presentation can thus lead to a fairly free interpretation of historical and architectural values. Therefore, the priority is to carry out all conservation activities consciously so that they meet the requirements guaranteeing the maintenance of the horizon of authenticity of the historical building [Petzet, 1995].

Given this, it should be noted that in striving to reflect the aspect of historicity, the historical character of a given home is often emphasized by the particular exposure of the oldest part. This, in consequence, leads to the frequent "extraction" of the original/primary parts and giving them the priority aesthetic value. This applies to both the exterior and interior of the building. Thus, later historical layers are removed (e.g. boarding or plastering), which, as part of the subsequent transformations of the building, have covered its original/primary fragments (in this case the parts of wooden log structures).

Photo 106. Benajah Gray house – intended exposure of the oldest, log part (photo T. Tomaszek, 2017).

These procedures, aimed at rendering the historicity of a given building, are often far-reaching author's interpretations, which threatens with the danger of serious falsifications the original message and almost any reading of the original solutions (Photo 107, 108, 109, 110). Consequently, secondary formal and material solutions may be unjustified. Therefore, we face the demanding task of maintaining and displaying the authentic historical values of a given object (and thus the cultural landscape of the whole area at the same time), at the same time with the necessity of maintaining the certificate of secondary changes and secondary building and architectural processes which in themselves constitute the horizon of authenticity of the historic building [Feilden, 1984].

Photo 107. Francis Waller house – maintenance of "old look character of the building" as interpretation of historicity (photo T. Tomaszek, 2017).

Photo 108. Francis Waller house – maintenance of "old look character of the building" as interpretation of historicity (photo T. Tomaszek, 2017).

Photo 109. Thomas Johnson house – maintenance of chosen character of the building as interpretation of historicity (photo T. Tomaszek, 2017).

Photo 110. Whitsett (Turner-Hill) house – rendering the historical character of the building through intended conservation activities – restored filling between logs emphasizing the feeling of the old age of structure (photo T. Tomaszek, 2017).

The efforts to guarantee the authentic formal and cultural message of the historic log buildings located in Cane Ridge therefore face many difficulties. Referring to the postulates generally accepted in contemporary conservation theory, included even in the Venice Charter, unquestionably should be assumed that the intended conservation actions should take into account the preservation of individual buildings together with their historical layers, which in themselves constitute the essence and character of these objects [Stipe, 1990], (Photo 111, 112).

Photo 111. B. Barnes house – authentic historical character of the building that is a result of "co-existence" of different architectural solutions from different time periods (photo T. Tomaszek, 2017).

Photo 112. Benajah Gray house – authentic historical character of the building that is a result of "co-existence" of different architectural solutions from different time periods (photo T. Tomaszek, 2017).

At the same time, it seems right to emphasize the demonstration and exposure of the oldest parts of buildings (i.e. wooden log structures), which from the point of view of the aspect of historicity constitute elements of extremely significant importance. However, the action of displaying the oldest part should take place in such a way and take place only in this case, when it is not threatening to falsify the formal and historical transmission of the building and the loss of its aspect of authenticity. So planned conservation activities/

actions will avoid therefore over-interpretation, which is often the case with historical buildings modeled on a specific, chosen epoch/period.

The reconnaissance made it possible to determine that basically in the examples of buildings from Cane Ridge containing in their structure original log remnants, noticeable exposition of the original/primary part of the structure is the result of the (historical) transformation processes in history (in time) and as such is not the result of only secondary modifying and "cleaning" activities to a given type of historical building. Thus, we can`t speak here about the situation of gross architectural or historical distortions, and thus the actions significantly weakening the horizon of authenticity.

At the same time, in the case of The Whitsett (Turner-Hill) house, we are dealing with an object where the log structure was secondarily completely screened by white board siding at some point in history and never deliberately exposed since. In this case, this secondary action is a historical solution and the consequence of a natural process of changes related to the existence of the building in time, and therefore contrary to the horizon of authenticity would be an action aiming to expose the original/primary parts of the building by the removal of white board siding. It is (white board siding) already the "organic" part of the building and thus has an equal historical, technological and aesthetical value (Photo 113).

Photo 113. Whitsett (Turner-Hill) house – the primary log structure later sheathed in white board siding (photo T. Tomaszek, 2017).

Another issue is the matter of possible, and often even evident, loss of historical character of a given building because of secondary operations or introduced elements. Of course, this matter remains debatable and its solution can take place on the technological, material or aesthetic level. Indeed, some secondary treatments, even if they are validated by the fact that they were applied historically, could at the same time lead to the lack of readability of original solutions, e.g. formal ones, by themselves not constituting a valuable addition/contribution. In this case, the removal of the effects of these activities or these elements seems to be fully justified. As an example, may be given some kind of balcony supported on the three pillars, obviously added later to the main body of the log structure of the Gillespie-Culbertson house, as well as the type and shape/colors of the window frames introduced into this log structure (Photo 114, 115, 116, 117). These elements, in addition to the falsification of the original formal solutions, significantly affect the weakening of the historical, and thus the authentic aesthetics of this building.

Photo 114. Gillespie-Culbertson house – some kind of balcony supported on the three pillars, obviously added later to the main body of the log structure, which is responsible for falsification of the original formal solutions and weakening historical, authentic aesthetics of this building (photo T. Tomaszek, 2017).

Photo 115. Gillespie-Culbertson house – some kind of balcony supported on the three pillars, obviously added later to the main body of the log structure, which is responsible for falsification of the original formal solutions and weakening historical, authentic aesthetics of this building (photo T. Tomaszek, 2017).

Photo 116. Gillespie-Culbertson house – the window frames introduced secondarily into the main body of the log structure, which are responsible for falsification of the original formal solutions and weakening historical, authentic aesthetics of this building (photo T. Tomaszek, 2017).

Photo 117. Gillespie-Culbertson house – the window frames introduced secondarily into the main body of the log structure, which are responsible for falsification of the original formal solutions and weakening historical, authentic aesthetics of this building (photo T. Tomaszek, 2017).

12   THE QUESTION OF AUTHENTICITY OF ORIGINAL TECHNIQUES AND TECHNOLOGICAL SOLUTIONS

Many factors are responsible for maintaining the horizon of authenticity of the historical building, which in themselves can be widely discussed. Nevertheless, it seems fully justified that to meet the requirement for the authenticity of a given building, it is legitimate to postulate the preservation of its original material and technological solutions, and likewise the application of original solutions and technologies in the process of its conservation and restoration [Larsen, Marstein, 2000]. Thus, with a view to analyzing buildings from Cane Ridge, it should be noted that this postulate has a reasonable reference to the examples discussed. Of course, the successive changes and transformations made on the buildings during their history have introduced new material and technological solutions, the validity of which (even they are secondary solutions to these used in the case of log structures) cannot be denied on new, secondarily introduced elements/parts[viii].

However, in the case of the original/primary parts which are "exposed" wooden log structures (or free-standing smaller log structures – eg. smokehouses), it is noticeable that the secondary materials and technologies were used incorrectly during subsequent renovations, which not only distorts the technological and material authenticity, but also the aesthetic authenticity of these objects. Therefore, in future conservation work, secondary and improper solutions as well as materials should be removed, and the original techniques and technologies should be used on the log structures. This applies for example mainly to fillings between individual logs, where traditional technologies[ix] have been replaced with modern ones (eg Portland Cement or polyurethane foam was used instead of traditional technologies[x]) (Photo 118, 119, 120, 121, 122, 123, 124, 125, 126, 127, 128, 129, 130, 131, 132, 133, 134, 135, 136).

Photo 118. Benajah Gray house – smokehouse, one of the two smaller log structures behind the house – noticeable remains of the original as well as secondary technologies in the layer of filling between the logs (photo T. Tomaszek, 2017).

Photo 119. Benajah Gray house – smokehouse, one of the two smaller log structures behind the house – noticeable remains of the original as well as secondary technologies in the layer of filling between the logs; detail (photo T. Tomaszek, 2017).

---

viii. These secondarily introduced elements/parts are the result of necessary changes that reflect the evolving uses of the buildings and needs of their owners, thus now there are historical elements in their own right. Therefore the materials and technologies introduced at these elements/parts (even they are not repeating primary solutions used in the case of the oldest parts – log structures) are original to them and the same their validity is fully justified.

ix. According to Weslager, the spaces between the logs forming the walls were traditionally filled with smaller pieces of wood or stones which were wedged tightly, and then caulked with moss, or wet clay, often tempered with animal hair or straw, in the process known as "daubing", or "chinking" (Weslager C. A. 1969. *The Log Cabin in America. From Pioneer Days to the Present.* Rutgers University Press, Quinn &Boden Company, New Brunswick, New Jersey, p. 15).

x. Very often introduced the new, secondary technologies are actually harmful to the log structures. The perfect example is the usage of Portland Cement. Its layer prevents the natural circulation of moisture and causes it [moisture] to deposit on the inside of the cement fillings, which leads to accelerated degradation of the wood tissue at these places (Ridout B. 1999. *Timber Decay in Buildings. The Conservation Approach to Treatment.* E.&F. Spon, London, UK)

Photo 120. Benajah Gray house – kitchen, second of the two smaller log structures behind the house – noticeable remains of the original as well as secondary technologies in the layer of filling between the logs (photo T. Tomaszek, 2017).

Photo 121. Benajah Gray house – kitchen, second of the two smaller log structures behind the house – noticeable remains of the original as well as secondary technologies in the layer of filling between the logs; detail (photo T. Tomaszek, 2017).

Photo 122. Francis Waller house – secondary technologies introduced in the layer of filling between the logs; eg. use of polyurethane foam (photo T. Tomaszek, 2017).

Photo 123. Francis Waller house – secondary technologies introduced in the layer of filling between the logs; eg. use of polyurethane foam – detail (photo T. Tomaszek, 2017).

Photo 124. Gillespie-Culbertson house – secondary technologies introduced in the layer of filling between the logs; eg. use of polyurethane foam – detail (photo T. Tomaszek, 2017).

Photo 125. Gillespie-Culbertson house – secondary technologies introduced in the layer of filling between the logs; eg. use of polyurethane foam – detail (photo T. Tomaszek, 2017).

Photo 126. Gillespie-Culbertson house – secondary technologies introduced in the layer of filling between the logs; eg. use of Portland Cement – detail (photo T. Tomaszek, 2017).

Photo 127. Ledbetter cabin – secondary technologies introduced in the layer of filling between the logs; eg. use of Portland Cement – detail (photo T. Tomaszek, 2017).

Photo 128. Ledbetter cabin – secondary technologies introduced in the layer of filling between the logs; eg. use of Portland Cement, bricks and pieces of metal – detail (photo T. Tomaszek, 2017).

Photo 129. Thomas Johnson house – secondary technologies introduced in the layer of filling between the logs; eg. use of Portland Cement (photo T. Tomaszek, 2017).

Photo 130. Thomas Johnson house – secondary technologies introduced in the layer of fillings of the losses of wood as well as the spaces between the logs; eg. use of Portland Cement – detail (photo T. Tomaszek, 2017).

Photo 131. Thomas Johnson house – secondary technologies introduced in the layer of filling between the logs; eg. use of Portland Cement – detail (photo T. Tomaszek, 2017).

Photo 132. Thomas Johnson house – secondary technologies introduced in the layer of filling between the logs; eg. use of Portland Cement – detail (photo T. Tomaszek, 2017).

Photo 133. Thomas Johnson house – remains of the original way of filling the spaces between the logs with small pieces of wood (photo T. Tomaszek, 2017).

Photo 134. Whitsett (Turner Hill) house – log smokehouse; secondary technologies introduced during the last conservation changing the authentic character of the structure (photo T. Tomaszek, 2017).

Photo 135. Whitsett (Turner Hill) house – log smokehouse; secondary technologies introduced in the layer of filling the spaces between the logs; eg. use of Portland Cement – detail (photo T. Tomaszek, 2017).

Photo 136. Whitsett (Turner Hill) house – log smokehouse; secondary technologies introduced during the last conservation changing the authentic character of the structure – detail (photo T. Tomaszek, 2017).

Due to the historical and technological multi-layering of the buildings, there is a need to maintain various, often contradictory technologies. Nevertheless, one should strive to maintain the right technology for a given historical and material part. This type of eclectic action is not only fully justified, but it also seems to be the best possible way to guarantee the authenticity of the buildings in question. It allows for the repetition and maintenance of solutions consistent with the era when given elements were created, which in is consistent with the principles of contemporary conservation thought/theory [Larsen, Marstein, 2000].

13 LOCATION AS AN ASPECT OF AUTHENTICITY

The majority of log structures surviving at Crane Ridge still stand where they were originally erected and thus meet the criterion of authenticity of the original location [Jokilehto, 1985]. An interesting exception, however, is the King cabin, which was moved from another place as part of rescuing it from demolition. The problem of transferring a wooden architectural object to a new place is always a matter of debate due to the issue of maintaining the horizon of the authenticity of this object [Larsen, Marstein, 2000]. In this case, we are dealing with a specific situation, because the Cabin was secondarily re-located on the land owned by the great-grandchildren of its builder. Of course, the spatial context has been changed – undoubtedly the cabin has a new location, although the same family context remained (it is still in the possession of the same family) and it still belongs to the same cultural landscape (Photo 137). Thus, it seems justified to postulate that despite the change of the physical place of location, the cabin maintains the authenticity of its original/primary level of meaning as a building owned and maintained by the same family.

Photo 137. King cabin (photo T. Tomaszek, 2017).

The question of the authenticity of the King cabin, however, remains an ambiguous matter. Along with the new location, the building gained a special dimension of historical significance, which was emphasized, for example, by the organization of the nearest surroundings – such as locating a historical wooden pole with a cast-iron bell[xi] next to the cabin structure (Photo 138, 139). This far-reaching interpretation established an additional message at the historical level, as well as creating a new emotional level for the immediate family. Despite the ideal insertion of this historical building into the spatial and cultural context of Cane Ridge, there is the danger of falsifying its original meaning by over-interpretation.

---

xi. Often bells like this were used to call people to meals, or to school.

Photo 138. King cabin and its surroundings (photo T. Tomaszek, 2017).

Photo 139. King cabin – historical wooden pole with an alarm bell next to the cabin structure (photo T. Tomaszek, 2017).

Additionally, evident adulteration occurred locally at the level of original technologies, even though the cabin is probably mostly of original material (Photo 140, 144). This is particularly true of the chinking between the individual logs, where partial secondary solutions were used, which contradicted the material and technological authenticity (Photo 141, 142). An example is the use of metal reinforcement in filling of the spaces between the logs in the lower parts of building (Photo 143). According to preliminary findings, the roof covering is also made using secondary materials and secondary technologies.

Photo 140. King cabin – detail – original wooden elements (photo T. Tomaszek, 2017).

Photo 141. King cabin – original wooden logs in contrary to the original technology filled in with secondary material in the spots of loss of wood (photo T. Tomaszek, 2017).

Photo 142. King cabin – detail – original wooden logs, in contrary to the original technology filled in with secondary material (now heavily degraded) in the spots of loss of wood – detail (photo T. Tomaszek, 2017).

Photo 143. King cabin – the use of elements of metal reinforcement in filling of the spaces between the logs in the lower parts of building (photo T. Tomaszek, 2017).

Photo 144. King cabin – detail – original wooden logs – the traces of original, traditional hand-processing of the surface of wooden logs (photo T. Tomaszek, 2017).

The issue of the authenticity of the transferred wooden building, raised in the case of the King cabin, also applies to the house located on the same property belonging to one of the great granddaughters of the cabin builder. The solid structure of that house is a kind of a cluster of two elements created at different times, wherein one of the elements, its wooden log part, is constructed of reused historical logs that used to be a building material of a warehouse located in Nashville. This shows the difficult nature of the issue of authenticity in the case of [re]construction of a wooden building using a secondary material, especially when this utilized secondary material is used for obtaining a completely different architectural form/ shape than that of the previous building [Petzet, 1995]. It seems, however, unquestionable that this house (located on the same property along with the King cabin) can't be treated as a wooden historical construction reflecting the original character of the building from which the material was reused, but only as a very casual narrative and historical interpretation (Photo 145).

Photo 145.   Home owned currently by Judith King Vulcano, located at the same property with King cabin – detail – the wooden log part constructed of material which was re-used from historical warehouse located in Nashville (photo T. Tomaszek, 2017).

Considering the original location as an attribute of authenticity of the historical building, it is worth mentioning another structure, namely the Thomas Johnson house (Photo 146, 147). According to historical data, this building (or at least its wooden log structure) as part of the conservation work carried out in the 1930s was completely dismantled and then reassembled in the same place. The practice of such conservation of wooden buildings is widely known in both the Western hemisphere and Asian countries [Charles, 1992]. Due to the non-durability of wood as a building material, this practice is widely recognized as compatible with the horizon of authenticity, as long as the building maintains the same form/shape, replicates original technological solutions and only degraded material is exchanged [Larsen, Marstein, 2000].

Photo 146.   Thomas Johnson house – the building completely dismantled and then reassembled in the same place during the conservation work carried out in the 1930s (photo T. Tomaszek, 2017).

Photo 147.   Thomas Johnson house – detail – the building completely dismantled and then reassembled in the same place during the conservation work carried out in the 1930s (photo T. Tomaszek, 2017).

A preliminary inspection of the Thomas Johnson house allows the postulation that it meets the above-mentioned requirements of the horizon of authenticity. However, related to the reconstruction there was inconsistent repetition of the original technologies (e.g. the way of filling the spaces between the logs). Whether the original solutions were falsified during conservation or whether they are the result of later and subsequent renovations remains a matter of clarification from future more in-depth analysis and research (Photo 148, 149, 150, 151, 152, 153, 154).

Photo 148.   Thomas Johnson house – noticeable inconsistent repetition of the original technologies during past repairs/conservations of the building (photo T. Tomaszek, 2017).

Photo 149.   Thomas Johnson house – secondary technologies introduced in the layer of filling between the logs; eg. use of Portland Cement – detail (photo T. Tomaszek, 2017).

Photo 150.   Thomas Johnson house – secondary technologies introduced in the layer of filling between the logs; eg. use of Portland Cement – detail (photo T. Tomaszek, 2017).

Photo 151.   Thomas Johnson house – noticeable inconsistent repetition of the original technologies during past repairs/conservations of the building (photo T. Tomaszek, 2017).

Photo 152. Thomas Johnson house – noticeable inconsistent repetition of the original technologies during past repairs/conservations of the building (photo T. Tomaszek, 2017).

Photo 153. Homas Johnson house – noticeable inconsistent repetition of the original technologies during past repairs/conservations of the building (photo T. Tomaszek, 2017).

Photo 154. Thomas Johnson house – remains of the original way of filling the spaces between the logs with small pieces of wood (photo T. Tomaszek, 2017).

14   THE AUTHENTICITY OF HISTORICITY IN RELATION TO THE ORIGINAL FORM AND AESTHETICS

The issue of authenticity of the historical building in the case of such structures as the Ledbetter cabin is a separate subject. This log structure is probably located in the same place where it was originally built. At the same time, the changes it underwent with time quite significantly altered the aesthetics of the building, despite the fact that it has retained a general form/shape almost identical to the original one. During its existence, the cabin was repeatedly re-modeled, which manifests itself in the diversity of secondary additives and technologies, and the complete loss of the original character of its interior. The result of all this is a rather specific conglomeration of form and material solutions that are not entirely compatible with each other (Photo 155, 156, 157, 158). The matter is complicated by the fact that the surface of the outer fragments of the wooden logs of the front wall are charred to a large extent because of the fire that took place in the latter half of the twentieth century (Photo 159, 160).

Photo 155. Ledbetter cabin – the diversity of secondary additives and technologies introduced during successive repairs/conservation works which significantly altered the original aesthetics of the building (photo T. Tomaszek, 2017).

Photo 156. Ledbetter cabin – the diversity of secondary additives and technologies introduced during successive repairs/conservation works which significantly altered the original aesthetics of the building – detail (photo T. Tomaszek, 2017).

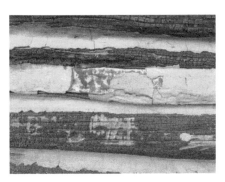

Photo 157. Ledbetter cabin – the diversity of secondary additives and technologies introduced during successive repairs/conservation works which significantly altered the original aesthetics of the building – detail (photo T. Tomaszek, 2017).

Photo 158. Ledbetter cabin – secondary technologies introduced in the layer of filling between the logs; eg. use of Portland Cement and pieces of metal – detail (photo T. Tomaszek, 2017).

As a building with a significantly adulterated original character, the Ledbetter cabin is simultaneously also an example of an object perfectly showing the free and uncontrolled historical changes of a wooden vernacular building. Although it would be reasonable to restore the cabin to its original appearance, on the other hand, this procedure will deprive it of a certain important context and thus a historical authenticity. Similarly, the situation of possible treatments concerning specific solutions is ambiguous, for example the method of conservation and exposure of the external front wall. Charred fragments of wooden elements themselves constitute a protective layer against degradation of wood tissue[xii], which is an argument for not exchanging/replacing them if the structural strength of these elements has not been significantly weakened. At the same time, the secondary effect obtained (dark charred surface), although authentic from the point of view of the historicity of the object, is incompatible with the original aesthetics and technology of the cabin.

Photo 159. Ledbetter cabin – the surface of the outer fragments of the wooden logs of the front wall charred to a large extent as a result of the fire that took place in the latter half of the twentieth century (photo T. Tomaszek, 2017).

Photo 160. Ledbetter cabin – the surface of the outer fragments of the wooden logs of the front wall charred to a large extent as a result of the fire that took place in the latter half of the twentieth century (photo T. Tomaszek, 2017).

15   CONCLUSIONS

The historic wooden log structures that have survived to this day in the Cane Ridge area are a perfect testimony to the vernacular architecture of the first settlers in Middle Tennessee. They are at the same time the essential component of the cultural landscape of these lands, reflecting the identity of the communities inhabiting it. The protection and preservation of these architectural objects for future generations is therefore of the utmost importance.

Selected significant examples of these buildings presented in the text show not only the richness of traditional solutions, but also the various fate of the later history of the

---

xii. The method of charring the surface of wooden elements to protect them from further degradation has been known since ancient times and was even applied to wooden poles for foundation in Venice (*Conservation of Historic Wooden Structures. Proceedings of the International Conference (Florence 22-27 February, 2005)*. 2005. Collegio degli ingegneri della Toscana, Firenze, Italy).

buildings during their existence. Proper understanding of the authenticity of these objects is the basis for identifying processes that have modified the original formal or technological solution, which is necessary to determine adequate conservation actions and at the same time for protecting them adequately in the future.

As historical objects of key importance for preserving the material and spiritual values of the early settlers of Tennessee, wooden log structures are at the same time carriers of architectural and aesthetic values that have shaped the present image of these lands. Without proper knowledge of these prototypes, it is not possible to fully understand the cultural processes that resulted in the contemporary landscape of the American South.

ACKNOWLEDGMENTS

The above research was carried out during the years of 2017/2018 under a Research Fellowship from the Center for Historic Preservation, Middle Tennessee State University, Murfreesboro, Tennessee, USA. It was also made possible thanks to a research grant from the Kosciuszko Foundation.

For their commitment and help in carrying out this research, special thanks goes to Professor Carroll Van West, PhD, Tennessee State Historian, Director of MTSU Center for Historic Preservation; Dr Stacey Graham, Research Professor at MTSU Center for Historic Preservation and Jenny Andrews, Historic Preservation Fellow at MTSU Center for Historic Preservation.

REFERENCES

"National Register of Historic Places Inventory – Nomination Form" for the Benajah Gray Log House. 1985. National Park Service, United States Department of the Interior.

Andrews J. 2017. Assessment *of Historic Structures in the Cane Ridge Community of Antioch, TN.*, MTSU Center for Historic Preservation, Murfreesboro.

Brown J. L. 1973. *Historic Cane Ridge and Its Families*. Blue&Gray Press, Nashville.

Charles F.W.B. 1992. *Dismantling, Repairing and Rebuilding as a Means of Conservation.*/in:/ ICOMOS UK: Timber Engineering Conference, Surrey University, Proceedings, ICOMOS UK, London.

Clements P. 1987. *A Past Remembered: A Collection of Antebellum Houses in Davidson County*. vol. I and II, edited and completed by Mason L. and Rogers S. T., Clearview Press, Nashville.

Conservation of Historic Wooden Structures. Proceedings of the International Conference (Florence 22-27 February, 2005). 2005. Collegio degli ingegneri della Toscana, Firenze, Italy.

Desch H. 1996. *Timber Structure, Properties, Convention and Use*. MacMillan, London, UK.

Feilden B. M. 1984. *A Possible Ethic for the Conservation of Timber Structures.*/in:/Charles F.W.B. *The Conservation of Timber Buildings*, Hutchinson, London, UK, pp. 238-241.

Jokilehto J. 1985. *Authenticity in Restoration* Principles *and* Practices. APT Bulletin 17, nr. 3-4: 5-11.

Larsen K. E., Marstein N. 2000. *Conservation of Historic Timber Structures. An ecological approach.* Butterworth-Heinemann Series in Conservation and Museology, Reed Educational and Professional Publishing Ltd., Great Britain.

McAlester V. S. 2015. *A Field Guide to American Houses: The Definitive Guide to Identifying and Understanding America's Domestic Architecture*. Knopf A. A., New York, USA.

McMurry S. 2001. *From Sugar Camps to Star Barns: Rural Life and Landscape in a Western Pennsylvania Community*. The Pennsylvania State University Press, University Park, Pennsylvania.

Petzet M. 1995. *"In the full richness of their authenticity"* – The Test of Authenticity and the New Cult *of Monuments.*/in:/Nara Conference on Authenticity/Conference de Nara sur l'Authenticite – Proceedings/Compte-rendu, Tapir Publishers/Agency for Cultural Affairs, Trondheim/ Tokyo, Norway/Japan, pp. 85-99.

Rehder J. B. 2012. *Tennessee log buildings: a folk tradition*. The University of Tennessee Press, Knoxville.

Ridout B. 1999. *Timber Decay in Buildings. The Conservation Approach to Treatment.* E.&F. Spon, London, UK.

Stipe R. E. 1990. *ICOMOS: A Quarter of a Century. Symposium sub-theme the Venice Charter.* (Prepared by US/ICOMOS),/in:/ICOMOS 9th General Assembly and International Symposium, Symposium Papers, Lausanne, ICOMOS Switzerland, pp. 407-424.

Weslager C. A. 1969. *The Log Cabin in America. From Pioneer Days to the Present.* Rutgers University Press, Quinn &Boden Company, New Brunswick, New Jersey.

MAPS

Beers D. G. 1878 *Map of Rutherford County, Tennessee: From New and Actual Surveys.* D. G. Beers&Co., Philadelphia.

Foster W. F. 1871. *Map of Davidson County Tennessee, from actual surveys made by order of the county court of Davidson County.* G.W.& C. B. Colton & Co., New York.

Reconstruction of a group of historic wooden buildings and the authenticity of the architectural heritage structure – a case study of Wynnewood, Tennessee

ABSTRACT: Reconstruction of a historical wooden building, usually resulting from the intention to preserve the cultural and historical identity of a given place, is a challenging conservation method regarding the issue of maintaining the authenticity of the building itself, as well as the authentic character of the historical site to which it belongs. This paper discusses this issue with an example of the reconstruction of a group of wooden historic buildings in Wynnewood, Tennessee. The site consists of six historical log buildings centered around a historic Sulphur mineral spring. It is an exceptional body of architecture reflecting the vernacular style of the frontier period in Southern American history. The biggest of the log buildings is the former inn, which is the largest standing log structure in Tennessee and most likely also throughout the entire USA. The reconstruction, which took place after a tornado destroyed much of the site, was at the same time a contemporary interpretation of a group of historical buildings. As such, even as it concerns the historical dimension and historical message, above all it was done on the architectural level and it was focused on the re-creation of the character of a place associated with the first settlers of Tennessee and on maintaining its authenticity and spatial cohesion.

1 INTRODUCTION

Wynnewood, sometimes also known as Castalian Springs, is a historic estate located on the west side of the hamlet of Castalian Springs (earlier known as Bledsoe's Lick) in Sumner County, Tennessee. It is about 34 miles northeast of Nashville. The site preserves a group of six original historical log buildings centered around the historic Sulphur mineral springs, thus it is an exceptional body of architecture reflecting the vernacular style of the frontier period. The biggest of the log buildings is the former inn (Photo 1). It is the largest standing log structure in Tennessee and most likely also throughout entire the USA[i] [Rehder, 2012, p.116]. The property was designated a National Historic Landmark in 1971 and has since been named "Wynnewood State Historic Site" and is administered by the state. Currently it is operated by the Bledsoe's Lick Historical Association[ii] together with the Tennessee Historical Commission.

---

i. According to *National Register of Historic Places Inventory-Nomination: Castalian Springs (Wynnewood)*, prepared by Morton III W. B., "The main house and dependencies constitute the finest remaining and most fully developed example of pioneer log architecture in the United States".

ii. Bledsoe's Lick Historical Association. 2013 – 2016. Wynnewood State Historic Site. http://bledsoeslick.com/(accessed August 2, 2018)

Photo 1.   Wynnewood State Historic Site (photo T. Tomaszek, 2018).

2   BRIEF HISTORY OF WYNNEWOOD

The architectural complex in Wynnewood was originally built as a stagecoach inn (for travelers between Nashville and Knoxville) and mineral springs resort/spa hotel in 1828 by three partners – Alfred Royal Wynne, Stephen R. Roberts and William Cage [Rehder, 2012, p. 123]. It was erected on land owned by Almira Winchester Wynne (wife of Alfred Wynne), who inherited 312 acres including the actual spring source from her father, Gen. James Winchester [Rehder, 2012, p. 123]. The name of the spring was changed from Bledsoe's Lick to Castalian Springs around 1830 by Wynne, due to a suggestion by his wife's brother, Valerius Publicola Winchester, who was referring to the Castalian Springs in Greek mythology (the spring which rose at the foot of Mount Parnassus) [Durham, 1974 (a), p. 140; Durham, 1994, p. 11]. After, the site was named "Wynnewood" in 1940 [Rehder, 2012, p. 117].

The stagecoach-inn business began to have serious problems in 1834 when the main east-west road, which to this point ran past Wynnewood, was moved about 12 miles to the

Photo 2.   Wynnewood State Historic Site in 1971 (photo: author unknown; source: NPS National Register of Historic Places Inventory – Nomination Form: Castalian Springs Wynnewood).

Photo 3.   Wynnewood State Historic Site in 1971 (photo: author unknown; source: NPS National Register of Historic Places Inventory – Nomination Form: Castalian Springs Wynnewood).

south. At that time Alfred Wynne bought shares from two other partners and became the sole owner of the property. He subsequently moved there with his family and lived there until his death in 1893.

Towards the end of the 19th century Wynnewood again served as a spa resort and was known as the "Castalian Springs House". At that time the resort was expanded and some of the new elements were added, including two rows of cottages at the back of the estate (about 100 feet behind the log inn), bowling alley, dance pavilion, poolroom and a large dining room [Durham 1974 (b), p. 315-316; Durham 1994, p. 50]. Unfortunately, only one cottage has survived from these elements.

At the turn of the 20th century, the site continued commercially as a Spa resort focused on the Springs [Rehder, 2012, p. 124] (Photo 4). By 1915, however, mineral springs were becoming less fashionable, thus the Wynne family were forced to close the business. They continued to work the farm and use the Inn as their home [Wynnewood State Historic Site. Undated]. The property remained in the hands of the family until 1971, when George Winchester Wynne, grandson of the house builder, donated the estate to the State of Tennessee as a historic site [Rehder, 2012, p. 124] (Photo 2, 3).

Photo 4.   "The Inn" in 1895 (photo: author unknown; source: Wynnewood State Historic Site. Undated. History. http://historicwynnewood.org/history; accessed September 10, 2018).

The Wynnewood property was hit by a powerful tornado (2008 Super Tuesday tornado outbreak[iii]) on the night of February 6, 2008. It severely damaged the architectural structures as well as century-old trees surrounding the place. The main building, the former inn, suffered major damages. It was shifted on its foundation, much of its upper floor was demolished and roof was ripped off [Rehder, 2012, p. 117].

The site was re-opened on July 4, 2012 after a four-year restoration project funded by, among others, the Federal Emergency Management Agency (FEMA), insurance proceeds, and the state government of Tennessee. The total cost of works was over $4 million.

3   ARCHITECTURE OF WYNNEWOOD

The former inn, which is the most important architectural structure in Wynnewood, is a building designed on the plan of an elongated rectangle. The size of the main compartment is approximately 110 feet (33,5 meters) long by 22 feet (6,7 meters) wide, and all of it is covered by one roof (Photo 5). However, when counting the detached kitchen, also a log unit which is connected to the rest by a breezeway, the building has length of 142 feet (43 meters) (Photo 6, 7).

Photo 5.   Wynnewood State Historic Site – the former Inn (photo T. Tomaszek, 2018).

Some sources describe the former inn as a structure built as an oversized version of the traditional dogtrot house, with two separate log structures joined via a central enclosed space under a common roof [Morton III, 1971]. Yet, according to Rehder, in this two-story structure two distinct, combined parts can be distinguished – a huge dogtrot part and

iii. The "2008 Super Tuesday tornado outbreak" was a deadly tornado outbreak which affected the Southern United States and the lower Ohio Valley from the afternoon of February 5 until the early morning of February 6. It is called "2008 Super Tuesday tornado outbreak" as the event began on Super Tuesday, while 24 United States were holding primary elections. The storm system produced several destructive tornadoes in heavily populated areas, most notably in the Memphis metropolitan area, in Jackson, Tennessee, and the northeastern end of the Nashville metropolitan area (Wikipedia, the Free Encyclopedia. 2019. 2008 Super Tuesday tornado outbreak. https://en.wikipedia.org/wiki/2008_Super_Tuesday_torna do_outbreak; accessed January 13, 2019).

Photo 6. Wynnewood State Historic Site – the detached kitchen (photo T. Tomaszek, 2018).

Photo 7. Wynnewood State Historic Site – the detached kitchen (photo T. Tomaszek, 2018).

a saddlebag part[iv]. Therefore, the main plan of the building cannot be considered to be based on one of them, as it is in reality a unique mix of these two types of design [Rehder, 2012, p. 118] (Figure 1). These are revealed/observed best on the north elevation (front one), with the dogtrot part nearer the east elevation and the saddlebag part to the west.

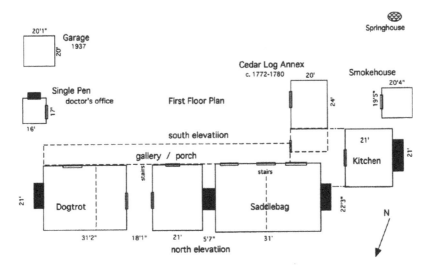

Figure 1. Distribution of individual architectural elements on the Wynnewood property – a schematic plan (based on sketch by Rehder; Rehder J. B. 2012. *Tennessee log buildings: a folk tradition*, The University of Tennessee Press, Knoxville, p. 120).

iv. Saddlebag plan – traditional form of cabin on the American frontier, mostly the southern frontier, which consisted of two rooms (or pens) with a central chimney that provided a hearth for each of the rooms. Each room had a door and perhaps a window or two and sometimes a connecting door between the rooms located on one side of the chimney.

Dogtrot plan – another traditional form of log house which consisted also of two rooms, but they were separated not by a wall, but by an open hallway, or dogtrot. Rooms and hallway shared a common floor and were covered by a common roof. Each room generally had a fireplace in its end wall and each had a door opening on the dogtrot. (Bealer A. W., Ellis J. O. 1978. *The Log Cabin. Homes of the North American Wilderness*. Barre Publishing, Barre, Massachusetts, p. 25)

And so, referring to Rehder, the north elevation can be divided and described in the following way (looking from east to west) (Photo 8):

1) an outside limestone chimney; (Photo 9)
2) the large east section of the dogtrot with six windows – three windows on each floor, wherein on the first floor are two rooms and on the second floor just one, big room
3) the dogtrot passage, which is open on the first floor (where there are the front steps and a porch) but enclosed on the second (but it has two windows there)
4) the west part of the dogtrot portion (which however at the same time makes the east portion of the saddlebag) with four windows, two on each floor
5) the enclosed, central chimney for the saddlebag
6) the west portion of the saddlebag section, which has six unevenly set windows (three on each floor containing for two rooms)
7) the west elevation's large limestone chimney [Rehder, 2012, p. 118-120] (Photo 10)

Photo 8.   Wynnewood State Historic Site – the former Inn (photo T. Tomaszek, 2018).

Photo 9.   Wynnewood State Historic Site – the former Inn (photo T. Tomaszek, 2018).

Photo 10.   Wynnewood State Historic Site – the former Inn (photo T. Tomaszek, 2018).

The main part of the former inn is two stories tall with 10 rooms under one roof (five rooms in a row on each floor). Therefore, on each of the two levels, the space can be described as one room deep by five rooms long. The first-floor plan is a 2-1-2 room layout while the second-floor plan is a 1-1-1-2 room layout. All rooms on the first floor either open onto the dogtrot or the gallery. Referring to Durham, all the rooms on the first floor were plastered in 1836, as were the western rooms on the second floor in later years. A large bunkhouse/meeting room located on the east end of the second floor is the only room in an unfinished state [Durham 1974 (a), p. 137; Durham 1994, p. 9].

The walls of the main building are constructed from 14 horizontal logs placed on each other, the lowest being the ground-level sill log. These logs are very large, up to 32 feet long (almost 10 meters), and they were squared to 16 inches wide (40 centimeters) by 8 inches thick (20 centimeters) [Rehder, 2012, p. 122]. The majority of the corner joints of logs are V notches. In the middle of the wall (exactly in the 8th log) mortified loft joists are fitted, creating both the ceiling for the first floor and the floor for the level of the second floor. The roof was originally covered with cedar shakes, although, according to historic data, in the 1950s they were replaced with a composite roof [Durham 1974 (a), p. 127; Durham 1994, p. 9].

The main entrance to the building is through the dogtrot passage, from where the stairs lead continuously to the rooms on the second floor and on the east end. A second set of stairs on the west end leads internally to the second floor.

On the south elevation there is a gallery which extends almost all the length of the building along the long axis of the main structure (Photo 11). The access to this gallery is through the doors leading to first-floor rooms. According to Rehder, in 1899 the gallery's overhang was lowered from just under the eaves to its present position halfway down so that it only shades the first floor [Rehder, 2012, p. 121]. On the south elevation there are also window openings, and they match the patterns of those on the north elevation. In turn, on the east and west elevation there are a large limestone chimneys with a width of over 7 feet (over 2 meters). Next to the west façade is located a one-story, free-standing log kitchen on a square plan of 21x21 feet (6.5x6.5 meters). The kitchen is separated from the main building by the covered passage. On the west façade of the kitchen is an extremely massive, 11-foot-wide (about 3.5 meters) limestone chimney (Photo 12, 13, 14, 15).

The oldest log structure in all the Wynnewood complex (sometimes called log annex) is located south of the kitchen (Photo 16, 17, 18). It is a two-story building which measures 20 by 24 feet (approximately 6x7,5 meters). It is unique as it is the only log structure at Wynnewood erected with cedar logs. During the days of the resort, it served as a dining room. Referring to Rehder, this log structure possibly predates the inn and could have been built by the initial settler in the area, Isaac Bledsoe, somewhere around 1772-1780 [Rehder, 2012, p. 121]. However, the other sources point out that this building, as a first

Photo 11. Wynnewood State Historic Site – gallery on the south elevation of the former Inn (photo T. Tomaszek, 2018).

Photo 12.   Wynnewood State Historic Site – former Inn, detached kitchen and smokehouse (photo T. Tomaszek, 2018).

Photo 13.   Wynnewood State Historic Site – extremely massive limestone chimney on the west facade of the detached kitchen (left) and smokehouse (right) (photo T. Tomaszek, 2018).

Photo 14.   Wynnewood State Historic Site – the detached kitchen separated from the main building by the covered passage (photo T. Tomaszek, 2018).

Photo 15.   Wynnewood State Historic Site – interior of the detached kitchen (photo T. Tomaszek, 2018).

Photo 16. Wynnewood State Historic Site – smokehouse (left) and cedar log annex (right) (photo T. Tomaszek, 2018).

Photo 17. Wynnewood State Historic Site – cedar log annex (left) (photo T. Tomaszek, 2018).

Photo 18.   Wynnewood State Historic Site – interior of the cedar log annex (photo T. Tomaszek, 2018).

structure on the Wynnewood site, was a two-story log cabin built by General James Winchester, Almira's father [Wynnewood State …, Undated, The Log…]. On the other hand, in the historical descriptions quoted by Durham, it was mentioned as being part of the estate of General James Winchester in 1829 and at that time probably performed the function of "comfortable cabin" or "workshop". [Durham 1974 (a), p. 138; Durham 1994, p. 10] In all likelihood, this building was connected to the main house by the board-and-batten covered passageway in 1899. Originally it had a stone fireplace and chimney which are now lost [Rehder, 2012, p. 121].

There are also other historical log structures located on the Wynnewood property, as, for example, the smokehouse and additional free-standing log house, a single-pen with large limestone chimney. The first one is standing west and south of the log annex and just a few feet from the southwest corner of the kitchen (Photo 19). It measures approximately 19 feet by 20 feet (approximately 5,8 by 6,1 meters). The second one, which is standing about 40 feet away to the southeast of the inn, measures 16 by 17 feet (approximately 4,9 by 5,2 meters) and it has a big chimney that is almost 6 feet (almost 2 meters) wide. According to historians, it was built at about the same time as the inn, and over the years it has played different functions (summer cottage, doctor's office and quarters, bachelor's quarters and school [Durham 1974 (a), p. 138; Durham 1994, p. 10]. Currently

Photo 19.   Wynnewood State Historic Site – view of the main group of buildings from the south-western side (photo T. Tomaszek, 2018).

it is converted to a doctor's office and it hosts the display of some medical artifacts used in the historical mineral springs resort.

4   TECHNOLOGICAL AND TECHNICAL SOLUTIONS

Most likely the Wynnewood complex was built of materials plentifully available on or around the site. The architectural structures were erected of sturdy hardwood logs that were all cut from nearby trees, forming the walls by setting the logs into place on the foundation. The wood was mainly white oak, walnut and ash. However, the smokehouse is almost entirely made of black walnut, and, as an exception, the log annex entirely from cedar[v] [Rehder, 2012, p. 122]. The foundation walls, as well as the original chimneys, were made of limestone blocks quarried from a nearby hillside. The roof, originally latticed with wide boards, was then covered with hand-split wood shingles [Wynnewood State ..., Undated, The Log...].

After pre-treatment with a saw, wooden logs at Wynnewood were processed by hand by means of woodworking tools commonly used in the first half of the nineteenth century in the United States. Some of these tools have survived and together with other historical tools[vi] are now stored in the attic of the kitchen (Photo 20, 21).

---

v.   According to "National Register of Historic Places Inventory-Nomination: Castalian Springs (Wynne-wood)", prepared by Morton III W. B., similarly like the smoke house, the original detached log kitchen with an exceptionally wide exterior stone chimney is also entirely made of black walnut, and apart from the log annex also the "new" kitchen, the office, and the garage are all made of cedar logs (the garage is a 20th century structure)

vi.   Some historical tools were used, inter alia, during the Wynnewood reconstruction

Photo 20. Wynnewood State Historic Site – historical tools stored in the attic of the kitchen (photo T. Tomaszek, 2018).

Photo 21. Wynnewood State Historic Site – historical tools stored in the attic of the kitchen (photo T. Tomaszek, 2018).

The final surface of the logs was obtained with the help of two tools – first the axe[vii] and then finally the adz[viii] (Photo 22, 23, 24, 25). The effect of using the latter

Photo 22. Wynnewood State Historic Site – axe and adz (historical tools stored in the attic of the kitchen) used originally for erecting and reconstruction of the architectural structures (photo T. Tomaszek, 2018).

Photo 23. Wynnewood State Historic Site – axe used originally for erecting and reconstruction of the architectural structures (photo T. Tomaszek, 2018).

vii. Axe, which is one of man's most ancient tools, was brought with the first immigrants from Europe and soon it became an essential carpenter's tool in New World. However its shape used so far proved to be inadequate to meet the needs of the cabin builders on the American southern and western frontiers. Thus, around 1740 the new type of axe was developed, called American long-handled broadaxe, designed especially for felling trees and building log cabins (Weslager C. A. 1969. *The Log Cabin in America. From Pioneer Days to the Present*. Rutgers University Press, Quinn &Boden Comp., New Brunswick, New Jersey, p. 11).

viii. Adz (properly called foot adz) was another basic hawing tool used in constructing a log structures and it was particularly used for finer hewing instead of the broadax. This tool has its origins in the Stone Age. It was actually "a chisel equipped with a handle so that the carefully sharpened edge could be swung to slice away the inequalities on top of a timber" (Bealer A. W., Ellis J. O. 1978. *The Log Cabin. Homes of the North American Wilderness*. Barre Publishing, Barre, Massachusetts, p. 35)

Photo 24. Wynnewood State Historic Site – the method of processing wooden logs by axe, used originally for erecting and reconstruction of the architectural structures (photo T. Tomaszek, 2018).

Photo 25. Wynnewood State Historic Site – the method of processing the surface of wooden logs by adz (visible vertical trace of the tool), used originally for erecting and reconstruction of the architectural structures (photo T. Tomaszek, 2018).

tool is typical for early log structures as the visual effect on the log's surface suggests the form of vertical incisions (Photo 26, 27).

Photo 26. Wynnewood State Historic Site – visual effect on the wooden logs surface in the form of vertical incisions (the effect of processing the surface by adz) (photo T. Tomaszek, 2018).

Photo 27. Wynnewood State Historic Site – visual effect on the wooden logs surface in the form of vertical incisions (the effect of processing the surface by adz) (photo T. Tomaszek, 2018).

To connect the logs three types of corner joints were used, all typical of the early log houses on the American frontier (V Notches, Diamond Notches and Half Dovetail Notches) (Photo 28, 29, 30). Although – as already mentioned – most corner joints are V notches.

The spaces between the logs forming the walls were originally filled using 19th century technology. Small pieces of wood (in some places also small stones) were wedged

Photo 28. Wynnewood State Historic Site – V Notches type of corner joints (photo T. Tomaszek, 2018).

Photo 29. Wynnewood State Historic Site – Diamond Notches type of corner joints (photo T. Tomaszek, 2018).

Photo 30. Wynnewood State Historic Site – Half Dovetail Notches type of corner joints (photo T. Tomaszek, 2018).

tightly, and then caulked with wet clay mixed with lime, sand and animal hair[ix] (Photo 31, 32).

---

ix. According to Weslager, this process of caulking is known as "daubing", or "chinking", and is the very traditional way of filling the spaces between the logs in early log houses in United States (Weslager C. A. 1969. *The Log Cabin in America. From Pioneer Days to the Present*. Rutgers University Press, Quinn &Boden Company, New Brunswick, New Jersey, p. 15)

Photo 31. Wynnewood State Historic Site – original way of filling the spaces between the logs with small pieces of wood (photo T. Tomaszek, 2018).

Photo 32. Wynnewood State Historic Site – original way of filling the spaces between the logs with small pieces of stones (instead of pieces of wood) (photo T. Tomaszek, 2018).

The original interior finish of most of the rooms with white wash was also in line with traditional methods used in the 19th century. Apart from the evident aesthetic dimension, this procedure was probably also a kind of wood surface protection (Photo 33, 34).

Photo 33. Wynnewood State Historic Site – remains of original interior finishing with white wash (photo T. Tomaszek, 2018).

Photo 34. Wynnewood State Historic Site – remains of original interior finishing with white wash (photo T. Tomaszek, 2018).

5   RECONSTRUCTION OF WYNNEWOOD

The destruction of architectural structures caused by a tornado was extremely advanced and extensive. The former inn building suffered exceptionally. The roof was totally broken off and the level of the second floor from the eastern side was completely devastated (Photo 35, 36, 37, 38, 39). Additionally, the building was slightly shifted on its foundations,

Photo 35. Wynnewood State Historic Site – damages of former Inn caused by a tornado (photo courtesy of Rick Hendrix).

Photo 36. Wynnewood State Historic Site – damages of former Inn caused by a tornado (photo courtesy of Rick Hendrix).

Photo 37. Wynnewood State Historic Site – damages of former Inn caused by a tornado (photo courtesy of Rick Hendrix).

Photo 38. Wynnewood State Historic Site – damage of the trees on the property caused by a tornado (photo courtesy of Rick Hendrix).

Photo 39. Wynnewood State Historic Site – damages of former Inn caused by a tornado (photo courtesy of Rick Hendrix).

which seriously weakened its structural stabilization and threatened further destruction, including the collapse of the remains of the surviving structure.

Due to the former inn's condition, comprehensive conservation and restoration works were required, which were carried out in several stages. The main goal of these activities was to restore both the whole complex and the former inn to the state before the tornado. Thus, the preservation of the character of the place and historical and cultural values associated with it became a priority. At the same time, emphasis was placed on the need to minimally distort the original architectural and technological solutions, which would guarantee the maintenance of the authenticity and historical coherence of this unique place.

Work began by securing the architectural structure with temporary roofing and recognizing the destruction status of individual elements "detached" from the main body of the building. On this basis, their structural durability was estimated, and it was determined which of them could still be used for reconstruction works. An important goal of this action was to preserve as many original elements as possible and thus reuse them in reconstructed parts (Photo 40).

Then, at the height of the upper boundary of the stone foundation (i.e. under the lowest log of the wall structure), the building was supported on a specially constructed structure made of metal rails and wooden beams. In addition, the whole structure was stabilized with vertical elements attached to external walls and embedded on metal rails. (Photo 41, 42, 43) This allowed the dismantling of the remains of the original foundation while maintaining stability of the entire structure, and then "shifting" the building to its original location/ embedding. Eventually, the foundation was reconstructed under the entire building, and the original material was used.

The next stage was the reconstruction of the level of the second floor of the former inn as well as the other destroyed architectural parts. The original elements broken off during the tornado which met the structural requirements were re-used, whereas that ones that were completely destroyed were replaced (Photo 44, 45, 46, 47). The reconstructed elements were made of the same species of wood and with similar properties. Importantly, the used wood was coming from trees growing on Wynnewood grounds and broken during a hurricane. The surface of new wooden elements introduced into the reconstructed walls was fabricated to imitate the original solutions/surface treated by hand. This effect, however, was obtained for the most part thanks to the use of modern tools/methods of machining, and not by using traditional tools.

Photo 40.   Wynnewood State Historic Site – selection of the wooden elements that could be re-used for reconstruction works (photo courtesy of Rick Hendrix).

Photo 41. Wynnewood State Historic Site – building of former Inn supported on a specially constructed structure made of metal rails and wooden beams (photo courtesy of Rick Hendrix).

Photo 42. Wynnewood State Historic Site – building of former Inn supported on a specially constructed structure made of metal rails and wooden beams (photo courtesy of Rick Hendrix).

Photo 43. Wynnewood State Historic Site – building of former Inn supported on a specially constructed structure made of metal rails and wooden beams; dismantling of the remains of the original foundation (photo courtesy of Rick Hendrix).

The next stage of works was the reconstruction of the destroyed external furnace chimney on the east façade of the former inn. Original material was used for this purpose and missing parts were supplemented from local sources of the same stone with similar properties. Nevertheless, the internal chimney shaft, which is invisible to observers, was made of a modern material – an airbrick (Photo 48, 49, 50, 51).

The roof supporting structure was then reconstructed (Photo 52, 54, 55, 56, 57). The roof truss hidden under the roof covering, although mapping the original shape, was made with a partial application of modern solutions, e.g. reinforcements of beam connections using modern screws (Photo 53). The roof covering finishing this stage of works was made in a traditional way. It is a layer of hand-made wooden shingles (Photo 58, 59).

The conservation and reconstruction works were carried out with other architectural elements of the Wynnewood complex (Photo 60, 61). In the end the object regained its splendor

Photo 44. Wynnewood State Historic Site – reconstruction of the level of the second floor of the former Inn (photo courtesy of Rick Hendrix).

Photo 45. Wynnewood State Historic Site – reconstruction of the level of the second floor of the former Inn (photo courtesy of Rick Hendrix).

Photo 46. Wynnewood State Historic Site – reconstruction of the level of the second floor of the former Inn (photo courtesy of Rick Hendrix).

Photo 47. Wynnewood State Historic Site – reconstruction of the level of the second floor of the former Inn (photo courtesy of Rick Hendrix).

Photo 48. Wynnewood State Historic Site – the reconstruction of the destroyed external furnace chimney on the east façade of the former Inn (photo courtesy of Rick Hendrix).

Photo 49. Wynnewood State Historic Site – the reconstruction of the destroyed external furnace chimney on the east façade of the former Inn (photo courtesy of Rick Hendrix).

Photo 50. Wynnewood State Historic Site – reconstruction of the level of the second floor of the former Inn (photo courtesy of Rick Hendrix).

Photo 51. Wynnewood State Historic Site – the reconstruction of the destroyed external furnace chimney on the east façade of the former Inn (photo courtesy of Rick Hendrix).

Photo 52. Wynnewood State Historic Site – the reconstruction of the roof supporting structure (photo courtesy of Rick Hendrix).

Photo 53. Wynnewood State Historic Site – the reconstruction of the roof supporting structure; reinforcements of beam connections using modern screws (photo courtesy of Rick Hendrix).

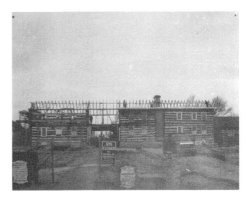

Photo 54. Wynnewood State Historic Site – the reconstruction of the roof supporting structure (photo courtesy of Rick Hendrix).

Photo 55. Wynnewood State Historic Site – the reconstruction of the roof supporting structure (photo courtesy of Rick Hendrix).

Photo 56. Wynnewood State Historic Site – the reconstruction of the roof supporting structure (photo courtesy of Rick Hendrix).

Photo 57. Wynnewood State Historic Site – the reconstruction of the roof supporting structure (photo courtesy of Rick Hendrix).

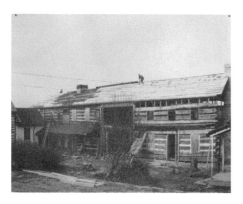

Photo 58. Wynnewood State Historic Site – the reconstruction of the roof covering – layer of hand-made wooden shingles (photo courtesy of Rick Hendrix).

Photo 59. Wynnewood State Historic Site – the reconstruction of the roof covering – layer of hand-made wooden shingles (photo courtesy of Rick Hendrix).

Photo 60. Wynnewood State Historic Site – the reconstruction works carried out on Cedar Log Annex (photo courtesy of Rick Hendrix).

Photo 61. Wynnewood State Historic Site – the reconstruction works carried out on the property (photo courtesy of Rick Hendrix).

that was lost during the tornado and its historical narration was recreated. At the same time, at first glance, any preservation and restoration works are virtually unnoticeable. Thus, it can be boldly said that the Wynnewood reconstruction may be considered a conservation success.

6 RECONSTRUCTION AND CONSERVATION OF WYNNEWOOD AS A CHALLENGE TO THE HORIZON OF THE AUTHENTICITY OF A GROUP OF VERNACULAR WOODEN HISTORIC BUILDINGS

Reconstruction and conservation of Wynnewood wooden architectural structures has been carried out with due care for details. Introduced new elements replacing damaged elements were appropriately merged with the remains of the original architectural body (Photo 62, 63, 64, 65, 66, 67). At the same time, traditional technological solutions were partly successfully used, maintaining the historic character of the place. We should discuss the processing of the new wooden elements (in the case of which the original method of surface processing was repeated) (Photo 68) and the filling method between the individual logs, which was carried out in accordance with the original/traditional solution. The gaps were filled with small flat stones placed at an angle, and then sealed through the process of caulking known as "daubing", with the composition of components repeating the original solutions (wet clay mixed with lime, sand and animal hair)[x] (Photo 69, 70, 71, 72).

It is worth mentioning that these traditional technological solutions were combined with modern technologies and conservation techniques, which can be criticized from the level of conservation methodologies aimed at maintaining original technologies (and thus aimed at

Photo 62. Wynnewood State Historic Site – the former Inn after conservation and reconstruction works with new elements replacing damaged elements and merged with the remains of the original architectural body (photo T. Tomaszek, 2018).

---

x. According to Weslager, this process of caulking is known as "daubing", or "chinking", and is the very traditional way of filling the spaces between the logs in early log houses in United States (Weslager C. A. 1969. *The Log Cabin in America. From Pioneer Days to the Present*. Rutgers University Press, Quinn &Boden Company, New Brunswick, New Jersey, p. 15)

Photo 63. Wynnewood State Historic Site – the former Inn after conservation and reconstruction works with new elements replacing damaged elements and merged with the remains of the original architectural body (photo T. Tomaszek, 2018).

Photo 64. Wynnewood State Historic Site – the former Inn after conservation and reconstruction works with new elements replacing damaged elements and merged with the remains of the original architectural body (photo T. Tomaszek, 2018).

Photo 65. Wynnewood State Historic Site – the former Inn after conservation and reconstruction works with new elements replacing damaged elements and merged with the remains of the original architectural body (photo T. Tomaszek, 2018).

Photo 66. Wynnewood State Historic Site – the former Inn after conservation and reconstruction works with new elements replacing damaged elements and merged with the remains of the original architectural body (photo T. Tomaszek, 2018).

Photo 67. Wynnewood State Historic Site – interior of the detached kitchen' attic after conservation and reconstruction works with new elements replacing damaged elements and merged with the remains of the original architectural body (photo T. Tomaszek, 2018).

Photo 68. Wynnewood State Historic Site – former Inn after conservation and reconstruction works with new elements replacing damaged elements, in case of which the original method of surface processing was used (photo T. Tomaszek, 2018).

Photo 69. Wynnewood State Historic Site – architectural structures after conservation and reconstruction works; introduced the traditional filling method of space between the individual logs (photo T. Tomaszek, 2018).

Photo 70. Wynnewood State Historic Site – architectural structures after conservation and reconstruction works; introduced the traditional filling method of space between the individual logs (photo T. Tomaszek, 2018).

Photo 71. Wynnewood State Historic Site – architectural structures after conservation and reconstruction works; introduced the traditional filling method of space between the individual logs (photo T. Tomaszek, 2018).

Photo 72. Wynnewood State Historic Site – architectural structures after conservation and reconstruction works; introduced the traditional filling method of space between the individual logs (photo T. Tomaszek, 2018).

maintaining the authenticity of the historic building in the traditional technological sense). It seems, however, that the modern solutions used did not disturb the overall original character of the building. Because they concern these operations, which do not represent the final aesthetics of the work done (such as pre-treatment of wood or materials used for concealed structural fragments – furnace interior shaft), the choice of modern building or technological methods did not significantly disrupt the essential historical or aesthetic message of the architectural complex of Wynnewood.

At the same time, the important action from the perspective of maintaining the horizon of authenticity of this unique group of log buildings was to leave the historical repairs and additions (Photo 73, 74, 75, 76, 77, 78), to introduce new additions in harmony with historical repairs, as well as to retain some naturally occurring deformations. An example of the latter may be leaving (left) deformation of the lower part of the Cedar Log Annex (Photo 79, 80). In addition, it was extremely valuable decision to preserve as many original elements as possible, which guaranteed (despite advanced reconstruction) maintaining the material authenti-

Photo 73. Wynnewood State Historic Site – Smokehouse after conservation and reconstruction works; visible remains of the previous (historical) conservation and repair works from the time before destruction caused by the tornado (photo T. Tomaszek, 2018).

Photo 74. Wynnewood State Historic Site – Smokehouse after conservation and reconstruction works; visible remains of the previous (historical) conservation and repair works from the time before destruction caused by the tornado – detail (photo T. Tomaszek, 2018).

Photo 75. Wynnewood State Historic Site – former Inn after conservation and reconstruction works; visible remains of the previous (historical) conservation and repair works from the time before destruction caused by the tornado (photo T. Tomaszek, 2018).

Photo 76. Wynnewood State Historic Site – former Inn after conservation and reconstruction works; visible remains of the previous (historical) conservation and repair works from the time before destruction caused by the tornado (photo T. Tomaszek, 2018).

Photo 77. Wynnewood State Historic Site – former Inn after conservation and reconstruction works; visible remains of the previous (historical) conservation and repair works from the time before destruction caused by the tornado (photo T. Tomaszek, 2018).

Photo 78. Wynnewood State Historic Site – Cedar Log Annex after conservation and reconstruction works; visible remains of the previous (historical) conservation and repair works from the time before destruction caused by the tornado (photo T. Tomaszek, 2018).

Photo 79. Wynnewood State Historic Site – Cedar Log Annex after conservation and reconstruction works; visible the naturally occurred in time, historical deformations (photo T. Tomaszek, 2018).

Photo 80. Wynnewood State Historic Site – Cedar Log Annex after conservation and reconstruction works; visible the naturally occurred in time, historical deformations (photo T. Tomaszek, 2018).

city of the objects as much as possible, and (with heavily patinated surface of majority of these surviving elements) transmission of the historical character of the buildings (Photo 81, 82, 83, 84, 85).

As a result, we are dealing with a group of buildings representing at the structural (formal) and material level the record of historical changes, and whose essence of authentic cultural and historical message, despite many modifications made, seems to remain undisturbed. At the same time, the individual buildings constitute a kind of compendium of knowledge about historical methods of conservation and methods of recognizing and displaying the historical values of the architectural structure. It is worth noting that these mentioned conservation interventions from the past – saved during the reconstruction of Wynnewood architectural structures regardless of their technological adequacy – show the richness of the methodologies used for preservation of the wooden log structure, although they can also be in parallel criticized from their level of validity or correctness (since they contradict the conservation as an action focused to maintain the authenticity understood as the preservation of original techniques and technologies).

Photo 81. Wynnewood State Historic Site – architectural structures after conservation and reconstruction works that successfully preserved the historical character of the buildings (photo T. Tomaszek, 2018).

Photo 82. Wynnewood State Historic Site – architectural structures after conservation and reconstruction works that successfully preserved the historical character of the buildings (photo T. Tomaszek, 2018).

Photo 83. Wynnewood State Historic Site – architectural structures after conservation and reconstruction works that successfully preserved the historical character of the buildings (photo T. Tomaszek, 2018).

Photo 84. Wynnewood State Historic Site – architectural structures after conservation and reconstruction works that successfully preserved the historical character of the buildings (photo T. Tomaszek, 2018).

Photo 85. Wynnewood State Historic Site – architectural structures after conservation and reconstruction works that successfully preserved the historical character of the buildings (photo T. Tomaszek, 2018).

As an example, we can point to the Portland cement fillings introduced during previous conservations, which not only contradict the authentic building technologies and thus the appropriate conservation solutions, but also pose a threat due to its properties (constituting a layer impermeable to humidity contributing to accelerated degradation of wood) (Photo 86, 87). Another example (in this case controversial due to, as well as, the plane of aesthetic interference and possible negative effect on the wood tissue, similar to fillings made of Portland cement) is saved, the historical filling of the losses in the wood layer with synthetic fiber used to repair fishing boats, fixed with a mixture of glue of unknown composition (Photo 88). Although one could postulate the removal of improper historical conservation procedures and replacing them with those that guarantee the maintenance of traditional technologies and techniques, maintaining this peculiar historical record seems to defend against criticism and complement the full picture recorded in this way of Wynnewood's preservation history.

These easy-to-read for a trained-eye historical conservation interventions, rightly criticized from the point of view of solutions aimed at maintaining the authenticity of historical technological solutions and often themselves being a contribution to further accelerated degradation of wood tissue, constitute nevertheless an authentic dimension of the historical group of wooden buildings that is Wynnewood. Authenticity understood in this way directly

Photo 86. Wynnewood State Historic Site – Portland cement fillings introduced during previous conservations (photo T. Tomaszek, 2018).

Photo 87. Wynnewood State Historic Site – Portland cement fillings introduced during previous conservations (photo T. Tomaszek, 2018).

Photo 88. Wynnewood State Historic Site – filling of the losses in the wood layer with the help of synthetic fiber used to repair fishing boats, fixed with mixture of glue of unknown composition (photo T. Tomaszek, 2018).

transmits the record of the historical and conservational existence of individual architectural structures. And although it may seem to be a contradiction, it is the testimony of past actions that falsify original technologies that ultimately gives the dimension of authentic meaning as a process of dynamics in the shaping of historical and conservation consciousness, and ultimately the cultural identity of the architectural structure.

7   SUMMARY

The reconstruction and conservation of Wynnewood, carried out after advanced destruction caused by a tornado, was a modern interpretation of a group of historical buildings. This interpretation mainly concerned the historical dimension and historical message, however, it materialized above all on the architectural level. And so, in the case of reconstruction of the historical context, it was focused on the re-creation of character of the place associated with the history of the first settlers of Tennessee and on maintaining of its historical cohesion.

The interpretation made in the case of Wynnewood in general passed its task. The visitor has the impression of entering a reality that has been stopped in time, and the later modifications are hardly recognizable at first glance. This is at the same time reflected in the level of applied interference in architectural structures, and in the method of made reconstruction, which, despite advanced conservation actions, guaranteed the transmission of a coherent formal context.

Though debatable from the point of proper conservation methodology, nevertheless, a particularly successful procedure was to leave a specific chronicle of historically used conservation procedures. At the cost of partial adulteration of traditional technological solutions, this decision had a positive effect on maintaining the horizon of authenticity of this unique place.

ACKNOWLEDGMENTS

The above research was carried out during the years of 2017/2018 under a Research Fellowship at the Center for Historic Preservation, Middle Tennessee State University, Murfreesboro, Tennessee, USA. It was also possible thanks to a research grant from the Kosciuszko Foundation.

Special thanks to Professor Carroll Van West, PhD; Tennessee State Historian, Director of MTSU Center for Historic Preservation; Stacey Graham, Ph.D.; Research Professor at MTSU Center for Historic Preservation and Rick Hendrix, Director, Wynnewood State Historic Site for their help and commitment in carrying out this research.

REFERENCES

Bealer A. W., Ellis J. O. 1978. *The Log Cabin. Homes of the North American Wilderness*. Barre Publishing, Barre. Massachusetts.
Cisco, J. G. 1909. *Historic Sumner County*. Tennessee, Nashville.
Durham W. T. 1974 (a). *Wynnewood*. Tennessee Historical Quarterly, 33, pp. 127-156.
Durham W. T. 1974 (b). *Wynnewood, Part II*. Tennessee Historical Quarterly, 33, pp. 297-321.
Durham W. T. 1974 (c). *Mexican War Letters to Wynnewood*. Tennessee Historical Quarterly, 33, pp. 389-409.
Durham W. T. 1974 (d). *Civil War Letters to Wynnewood*. Tennessee Historical Quarterly, 34, pp. 32-47.
Durham W. T. 1994. *Wynnewood: Bledsoe's Castalian Springs, Tennessee*. Castalian Springs, TN: Bledsoe's Lick Historical Society.

Emergency Repairs&Restoration Wynnewood State Historic Site, SBC Project No. 160/004-01-2008-01. Project blueprints and documentation. Centric Architecture, Nashville, Tennessee.

Library of Congress. Undated. Prints & Photographs Reading Room, Prints & Photographs Online Catalog. Wynnewood, Gallatin-Hartsville Pike (State Highway 25), Castalian Springs, Sumner County, TN. https://loc.gov/pictures/item/tn0152/(Accessed October 4, 2018).

Morton III W. B. 1971. *National Register of Historic Places Inventory-Nomination: Castalian Springs (Wynnewood)*, Office of Archeology and Historic Preservation, History Division, Historic Sites Survey, National Park Service (available at: https://npgallery.nps.gov/NRHP/GetAsset/NHLS/71000838_text).

National Park Service, U.S. Department of the Interior. Undated. NPGallery Digital Asset Management System. National Register of Historic Places Inventory – Nomination Form: Castalian Springs (Wynnewood). https://npgallery.nps.gov/AssetDetail/NRIS/71000838 (accessed October 13, 2018).

National Park Service, U.S. Department of the Interior. Undated. National Register of Historic Places Inventory – Nomination Form: Castalian Springs (Wynnewood). https://npgallery.nps.gov/GetAsset/729db15c-4732-4df3-beac-b5473f50c2f0/(accessed October 13, 2018).

Project Manual, September 15, 2009; Emergency Repairs&Restoration Wynnewood State Historic Site, Castalian Springs, Tennessee; SBC Project No. 160/004-01-2008-01, Centric Project #08018; Centric Architecture, Nashville, Tennessee.

Rehder J. B. 2012. *Tennessee log buildings: a folk tradition.* The University of Tennessee Press, Knoxville.

Weslager C. A. 1969. *The Log Cabin in America. From Pioneer Days to the Present.* Rutgers University Press, Quinn &Boden Comp., New Brunswick, New Jersey.

West Van C. 1995. *Tennessee's Historic Landscapes: A Traveler's Guide.* Knoxville, University of Tennessee Press.

Wynnewood State Historic Site. Undated. History. http://historicwynnewood.org/history (accessed September 10, 2018).

Wynnewood State Historic Site. Undated. The Log Inn. http://historicwynnewood.org/the-log-inn (accessed October 20, 2018).

Authenticity versus interpretation – issues of the preservation of historical wooden buildings using the example of The Tipton-Haynes Historic Site and The Historic Sam Davis Home and Plantation, significant historic farms in Tennessee.

ABSTRACT: Preservation, which also includes the interpretation of groups of historical wooden buildings, often leads to modifications of the original character and narrative of these places, thus the loss of their authenticity. This paper discusses – in the perspective of the analysis of the issue of maintaining the authenticity of the wooden historical structure – two significant historical farms with outbuildings located in Tennessee – The Tipton-Haynes Historic Site and The Historic Sam Davis Home and Plantation. Both sites represent typical spatial and architectural layouts of farms from the beginning of the 19th century, both preserve typical types of historical farms outbuildings which in the majority are log structures, and finally, both are also architectural complexes in historical settings. While the example of The Tipton-Haynes Historic Site shows the variety of ways of preserving and interpreting a wooden building within a group of wooden structures located in one heritage site and the resulting issues of maintaining authenticity, The Historic Sam Davis Home and Plantation is a historical site where secondary technologies and conservation methods can be identified and, at the same time, it is a place in which one can critically discuss the issue of the translocation of wooden buildings into a new context as a part of the process of site interpretation.

1 INTRODUCTION

The groups of wooden buildings on historic farms in the southern United States are an important testimony of the cultural landscape of these lands. Their protection, interpretation and proper exposure encounter many problems, which often affect the modification of the original character and historical narrative of these places. The article discusses – in the perspective of the analysis of the issue of maintaining the authenticity of the wooden historical structure – the two significant historical farms with outbuildings located in Tennessee – The Tipton-Haynes Historic Site and The Historic Sam Davis Home and Plantation.

Both sites represent typical spatial and architectural layouts of farms from the beginning of the 19th century, they have the typical residential houses of the era, with similar histories of later transformations, and they present typical types of historical farm outbuildings, most of which are log structures. Both are also architectural complexes in a historical setting.

The history of both farms also begins similarly, namely from the construction of a log cabin (a typical pioneer's dwelling for the American frontier) which, while remaining the basic and essential part of the family home for future generations, at the same time undergoes in subsequent epochs the modifications that were typical for similar buildings in Tennessee. And so, among the most important changes in both cases, the changes executed in the Greek-Revival style determine the final form of the building. Most importantly, however, apart from witnessing similar historical changes, both places were eventually created as a historic sites, and both carry interpretations of a specific historical context and historical group of the wooden buildings, including interpretations of slavery. Thus, both places have log slave cabins and other typical farm outbuildings.

106

The example of The Tipton-Haynes Historic Site shows the variety of ways of preserving and interpreting a wooden building within a group of wooden structures located in one heritage site, and the resulting issues of authenticity. The Historic Sam Davis Home and Plantation is a perfect example of a place where secondary technologies and conservation methods can be analyzed, as well as the issue of the translocation of a wooden building into a new context as a part of the process of interpretation of a historical place.

2   THE TIPTON-HAYNES HISTORIC SITE

The Tipton-Haynes Historic Site is a property with eleven historical architectural wooden structures. It sits in the southwest portion of Johnson City in Washington County, Tennessee. It is managed by The Tipton-Haynes Historical Association, formed in 1965 [Tipton Haynes, National Register..., 1970]. The site consists of multiple types of buildings: the Main House (Photo 1), the law office, and various outbuildings included in the domestic complex behind the main house – the loom house, the necessary, smoke house, sorghum shed and a log cabin interpreted as the George Haynes slave cabin. Also at the site are a double crib log barn, a pigsty, a corncrib, a still house, and a springhouse [Van West, Gavin, Gardner, 2011, p. 15].

Photo 1. Front elevation of the Main House at the Tipton-Haynes Historic Site (photo T. Tomaszek, 2018).

3   HISTORY OF THE TIPTON-HAYNES HISTORIC SITE/HOUSE

The Tipton-Haynes property is considered one of the most historic sites in Tennessee. The story of this special place begins in 1784, when Colonel John Tipton from Virginia came to East Tennessee, known at that time as the Tennessee Country of North Carolina [Williams, 1970, p. 326], and purchased land from Samuel Henry on May 15 that year for the sum of fifty pounds North Carolina currency [Deed from Samuel Henry...., 1784, p. 301]. On his new property near Sinking Creek he soon built a two-story log cabin [Lawson, 1970, p. 105], which was the first architecture structure existing in long history of that place [Van West, Gavin, Gardner, 2011, p. 3].

Colonel Tipton lived in his home until his death in 1813. He was buried in the Tipton-Haynes Cemetery, which is located at the property. As he died intestate, the farm was inherited in 1813 by one of his sons, John Tipton, Jr. who managed the estate until his death in 1831 while serving in the General Assembly in Nashville [Van West, Gavin, Gardner, 2011, p. 5]. In 1837 the farm was sold by three of the heirs, Samuel P. Tipton, Elizabeth Tipton, and Edna Tipton to David Haines of Carter County (for the sum of $1,050.26) [Deed from Samuel P. Tipton, Elizabeth Tipton...., 1837, p. 177], who handed land together with Tipton's home as a wedding present to his son Landon Carter Haynes (who was born in Carter County, Tennessee in 1816 as the son of David Haynes and Rhoda Taylor) [Thomas, 1989, p. 57], [Van West, Gavin, Gardner, 2011, p. 5].

On July 1, 1865 Landon Haynes sold belonging to him the two hundred fifty-acre farm to John R. Banner [Van West, Gavin, Gardner, 2011, p. 7]. In 1867, John Banner then sold the property to Robert Haynes, the son of Landon C. Haynes [Deed from John R. Banner to Robert W. Haynes..., p. 499].

The next owner of the property was John White, who purchased the farm at auction on February 25, 1871. Next, in 1872, he conveyed the 220-acre property to Sarah L. Simerly, a niece of Landon C. Haynes [Deed from John White to S.L. Simerly..., p. 316] and wife of Samuel W. Simerly of Carter County [Van West, Gavin, Gardner, 2011, p. 7].The Simerly family continued to work the farm until the death of the elder Samuel W. Simerly on January 6, 1888 (they also have a son called Samuel W. Simerly, Jr.). Widowed Sarah L. Simerly lived with her sons in the house until her death in 1935 [Van West, Gavin, Gardner, 2011, p. 8] (Photo 2).

Photo 2.   The Main House during the Simerly period of ownership (photo W. Jeter Eason, June 12, 1936, the Historic American Buildings Survey Collection, Library of Congress, image courtesy the Center for Historic Preservation, Middle Tennessee State University, Murfreesboro, Tennessee).

Thanks to commitment of Judge Samuel C. Williams, the Chairman of the Tennessee Historical Commission, who felt that the Tipton-Haynes property was one of the most historically important sites in Tennessee [Van West, Gavin, Gardner, 2011, p. 11] the State purchased the house and 17.5 acres from the Simerly brothers on November 25, 1944 (for $7,552.25) [Deed from Samuel W. Simerly..., 1944, p. 344]. At that time a Memorandum was signed between the parties that outlined the details of the transaction. During the lifetime of the Simerly brothers, the State would have the right to enter the property to do some gardening works, but at the same time the Simerly brothers reserved the exclusive right to use the main house and all the remaining buildings.

In 1958 the Simerly brothers sold their remaining land to Clinton Garland (a neighbor and developer) and his wife [Van West, Gavin, Gardner, 2011, p. 10]. After the death of the last of the brothers in 1962, the State came into full ownership of the property. At the same

time, Clinton Garland received all farming tools, livestock, hay, feed, and a lot on Chero-
kee Road in Johnson City – all of it was the part of the property sold to him in 1958 [Van
West, Gavin, Gardner, 2011, p. 10].

For a number of years the site was not used while various parties discussed how to best
preserve and restore the property. They had also some conflicting concepts as to what period
the main house should be restored [Van West, Gavin, Gardner, 2011, p. 13]. One of the
voices was that of Charles W. Warterfield, Jr., AIA, an architect from Nashville, Tennessee,
who visited the site in 1967, when some of the restoration works had been started. His idea
was to follow the method of restoration done on Travelers Rest in Nashville. Like Tipton-
Haynes, Travellers Rest was a historical house with a series of additions that dated to different time and style periods. In order to deal with all of it, Travellers Rest is preserved in such
a way, that the each section was restored to its own date of origin or kept in its existing condition, depending on the particular factors for that individual section [Waterfield, 1967].

When the buildings were being restored, a new idea was to turn the site into a living
farm. At that time, on June 23, 1970 the local newspaper *Johnson City Press Chronicle*
reported, that "at the farm they would be bred cows, horses, sheep, goats and ducks, as
well as they would have a cane patch and a patch of tobacco" [Tipton-Haynes „living
farm"...., 1970]c Additionally, it was decided that some family who are caretakers of the
property would work on the farm and grow crops common to the area in the eighteenth
century. All of it was organized to interpret the lives of pioneer through demonstrations
[Van West, Gavin, Gardner, 2011, p. 14].

The site was known as the Tipton-Haynes Living Farm and existed in that form until
1986, when it was changed to Tipton-Haynes Farm to meet the city regulations that did
not allow farm animals to live within the city limits. Apart from using the property as
a living farm there were also the plans in the 1970s for building an amphitheater (that
would seat 1,500 to 1,800 visitors) and hosting outdoor dramas. However, this structure
was never constructed [Van West, Gavin, Gardner, 2011, p. 14].

The Tipton-Haynes Historic Site has been open to the public since April 17, 1971. It is
interpreting the history of the Tipton and Haynes families that once called this property
their home [Van West, Gavin, Gardner, 2011, p. 14]. After the State purchased an additional 27.6 acres in 2001, it consists of approximately forty-four acres.

4   THE TIPTON-HAYNES HISTORIC SITE – DESCRIPTION OF THE MAIN
HOUSE AND THE OUTBUILDINGS

The main house

The main house can be described as a two-story white frame house with a gable roof and
a porch of Greek Revival style. Its primary version, a two-story log cabin with a cellar, was
built by Colonel John Tipton in circa 1784. The foundation and a two-story stone chimney
were made of fieldstone. In the rear of the cabin there was probably a log kitchen with
a half-stone, half-log chimney [Tipton Haynes, National Register..., 1970]. According to
Ray Stahl, there is information left by General Thomas Love, a contemporary of Colonel
Tipton, who described the log cabin "as a large-size house, 25 by 30 feet, hewed logs,
a story and a half, no windows below – two or three window holes round in each gable and
above – a door in the front." [Stahl, 1986, p. 226]. Later, in 1798, Colonel John Tipton
erected on the same site a new log cabin that had the same footprint as the previous one
[Van West, Gavin, Gardner, 2011, p. 17].

The house underwent some changes when it was enlarged, most likely in the 1830s. The
north wall of the cabin was removed, and a frame structure was introduced, and all was
covered by weatherboarding. Also, new entrances to the front and back were constructed
[Lawson, 1970, p. 117]. However, the most significant alteration to the house took place during

the ownership of Landon Carter Haynes (1838-c. 1870), who added an ell to the rear (Photo 5), and introduced a dining room, a kitchen and a side room adjacent to the dining room. He also replaced the front porch with one in the Greek Revival style [Baratte, 1970, p. 125]. At that time, the house transformed from a hall and parlor style building to its current form as a central hall house, the plan that was predominant in Tennessee architecture during the 1810-1850 period [James, 1981, p. 63], [Van West, Gavin, Gardner, 2011, p. 17] (Figure 1)

Photo 3. The Main House in 1948 – photo showing that there were no shutters on the windows at that time (author unknown, the Tennessee Historical Society Picture Collection, Tennessee State Library and Archives, image courtesy the Center for Historic Preservation, Middle Tennessee State University, Murfreesboro, Tennessee).

Photo 4. The Main House in 1947 – photo showing brick cladding over the porch foundation (author unknown, Tennessee Department of Conservation Photograph Collection, Tennessee State Library and Archives; image courtesy the Center for Historic Preservation, Middle Tennessee State University, Murfreesboro, Tennessee).

The wide-ranging conservation and renovation works were done in 1965, after the State gained full possession of the house in the early 1960s [Van West, Gavin, Gardner, 2011, p. 17]. It included several changes. Among others, the chimneys wore torn down and replaced, a door was cut in the dining room, replacing a window in the ell of the house, the brick fireplaces and hearths were restored, the front porch was replaced with a different one, the square posts on the ell porch were replaced with similar which were used on the front porch [Specifications for restoration..., undated]. The house was subsequently restored in 1976, 1980s, 1985 and 1986 [Van West, Gavin, Gardner, 2011, p. 17]. Then in the period of years 1990 – 1991 it underwent succeeding restoration works that were focused on bringing the house to its appearance from the Haynes period of ownership [Van West, Gavin, Gardner, 2011, p. 18].

Although the current house no longer has the character of a log building, its structure retains the original fragment of the log cabin. As the building underwent advanced formal changes and was enlarged, the former outer south wall of the cabin is now enclosed by later additions and so it makes the interior wall between the old part of the house and the enclosed porch (Figure 1). The latter (enclosed porch) runs the full length of the original house and a portion of the original wall from the Tipton-era log cabin is exposed in this room on its north side (Photo 6). The rest of the room has vertical planking on the walls. The space in the enclosed porch is currently used for housing looms, spinning wheels, and other artifacts associated with weaving [Van West, Gavin, Gardner, 2011, p. 42]. It was organized in that way to give some interpretation of the architecture of the original cabin as well as usage of the later built addition of enclosed porch.

The exposed log section of the wall of the original house (which is made of chestnut wood) is in relatively in good condition. On its part located on the north side of enclosed

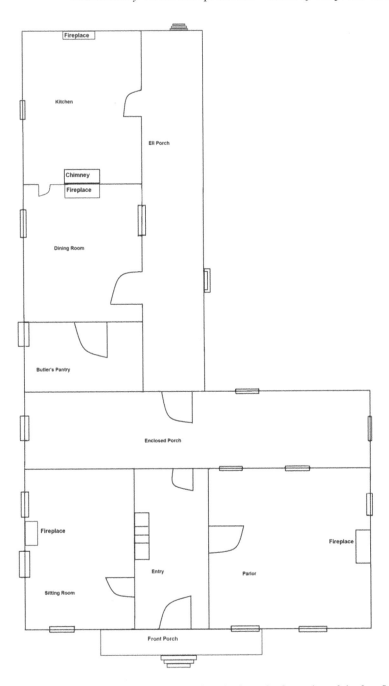

Figure 1. The main house at the Tipton-Haynes Historic Site – the floor plan of the first floor (drawing courtesy the Center for Historic Preservation, Middle Tennessee State University, Murfreesboro, Tennessee).

Photo 5. The main house at the Tipton-Haynes Historic Site – north elevation where it adjoins the ell addition (author and date unknown, image courtesy the Center for Historic Preservation, Middle Tennessee State University, Murfreesboro, Tennessee).

Photo 6. The main house at the Tipton-Haynes Historic Site – the west end of enclosed porch (author and date unknown, image courtesy the Center for Historic Preservation, Middle Tennessee State University, Murfreesboro, Tennessee).

porch the daubing between the logs was restored, and it was done in general in accordance with the original technology [Van West, Gavin, Gardner, 2011, p. 102]. The situation looks differently on the other parts, currently covered by planking, where the losses in chinking and daubing are noticeable (Photo 7).

In addition to the changes to the house, there were once more outbuildings associated with the site. *An Inventory of State Land*, dated November 24, 1960, notes that there was

Photo 7. The remained wall of the original log cabin – the fragment on the south elevation of the main house, covered currently by planking; noticeable the advance losses in chinking and daubing (photo T. Tomaszek, 2018).

a well house to the right rear (northeast) of the house as well as a small tool house and smokehouse to the left side rear (northwest) of the house [Tipton-Haynes Historic Site: Correspondence. ..., 1960].

The outbuildings of the farm

Apart from a few frame buildings, there are seven log structures on the property (Figure 2). These are domestic outbuildings (George Haynes slave cabin and smokehouse), as well as farm outbuildings (double crib log barn, corn crib, pigsty, still house and spring house). They constitute architectural elements of the interpretation

TIPTON-HAYNES HISTORIC SITE
SELF-GUIDED WALK

Figure 2.   The schematic map of at the Tipton-Haynes Historic Site (image courtesy the Center for Historic Preservation, Middle Tennessee State University, Murfreesboro, Tennessee).

of historical space as well as the interpretation of the forms/shapes and technologies of historical buildings. Due to the multifaceted nature of the above narrative, they simultaneously show the issue of the horizon of authenticity of heritage place/building and its application in this case.

Smokehouse

The smokehouse is located north of the main house. The logs have been chinked and daubed. It has a stone pier foundation and a roof is covered by wooden shingles (Photo 8). In the interior is an earthen floor and exposed ceiling beams (Photo 9, 10). The building is not currently used as a smokehouse, but is used to store some miscellaneous landscaping items [Van West, Gavin, Gardner, 2011, p. 61].

Photo 8.   Smokehouse at the Tipton-Haynes Historic Site (image courtesy the Center for Historic Preservation, Middle Tennessee State University, Murfreesboro, Tennessee).

Photo 9. Smokehouse at the Tipton-Haynes Historic Site – interior (image courtesy the Center for Historic Preservation, Middle Tennessee State University, Murfreesboro, Tennessee).

Photo 10. Smokehouse at the Tipton-Haynes Historic Site – interior (image courtesy the Center for Historic Preservation, Middle Tennessee State University, Murfreesboro, Tennessee).

This wooden log structure was built in 1967 during the period of the Tipton-Haynes restoration [Tipton-Haynes Historical Site: Restoration. ..., 1966-1967] and it was the first building reproduced at that time. It was erected according a traditional manner of construction and design [Baratte, 1970, p. 128] and traditional technological solutions were used [Van West, Gavin, Gardner, 2011, p. 61] (Photo 11). The logs were cut, fabricated and hand-processed in a traditional way, which allowed for the historical character of the surface of the wooden elements (e.g. typical for manual machining vertical cuts on logs) (Photo 13, 14). Also, for the filling the spaces between the logs the traditional methods was used, and the chinking/daubing was executed according to historical technologies (Photo 12).

Photo 11. Smokehouse – reconstruction of traditional covering of the roof made of shingles (image courtesy the Center for Historic Preservation, Middle Tennessee State University, Murfreesboro, Tennessee).

Photo 12. Smokehouse – reconstruction of traditional covering of the roof made of shingles, the daubing made during reconstruction according to historical technologies (image courtesy the Center for Historic Preservation, Middle Tennessee State University, Murfreesboro, Tennessee).

Photo 13. Smokehouse – the surface of the logs showing the traditional way of hand-processing (image courtesy the Center for Historic Preservation, Middle Tennessee State University, Murfreesboro, Tennessee).

Photo 14. Smokehouse – the surface of the logs showing the traditional way of hand-processing (image courtesy the Center for Historic Preservation, Middle Tennessee State University, Murfreesboro, Tennessee).

George haynes slave cabin

The cabin is located northeast of the smokehouse. It is a log structure set on a stone pier foundation and its roof is covered by wood shingles (Photo 15, 16, 17). It has a wooden porch on the front with a shed roof, square wooden posts, and a wooden floor (Photo 18). The front elevation has two bays, one window and one single leaf wooden door. The structure has also an exterior stone and brick chimney (Photo 19) [Van West, Gavin, Gardner, 2011, p. 64].

In its interior the cabin contains one room on the lower level and a loft area above, with a ladder (Photo 20, 22). The ceilings are wood beams, and the floors are made from wide wood planks. A stone fireplace with a mantel is located on the structure's south wall (Photo 21) [Van West, Gavin, Gardner, 2011, p. 66].

This log structure was located secondarily on the property of the Tipton-Haynes Historic Site. Originally it was erected in the Boones Creek area of Washington County circa 1840 by Henry and Elizabeth Fox as a cabin with a loft [Tipton-Haynes Historic Site Vertical Files]. In 1999, the building was donated to the Tipton-Haynes Historic Site by Stuart Wood and then moved there to aid in interpreting slavery at the site (Photo 23, 24). The work of dismantling the Fox cabin at its location and moving to Tipton-Haynes was performed by Leatherwood, Inc. of Fairview, Tennessee. After transfer to the site, the cabin was then reassembled. However, the Fox cabin no longer had a chimney, so the staff began a search. This task was soon fulfilled, as two chimneys from abandoned home sites in Greene County, Tennessee were donated by Kenneth Jenkins of Bulls Gap, Tennessee to Tipton-Haynes Historic Site. These chimneys were then used to reconstruct the chimney currently at the George Haynes Slave Cabin [Van West, Gavin, Gardner, 2011, p. 65].

Photo 15. George Haynes Slave Cabin at the Tipton-Haynes Historic Site – front elevation (photo T. Tomaszek, 2018).

Photo 16. George Haynes Slave Cabin at the Tipton-Haynes Historic Site – rear elevation (photo T. Tomaszek, 2018).

Photo 17. George Haynes Slave Cabin at the Tipton-Haynes Historic Site – detail: stone pier foundation (image courtesy the Center for Historic Preservation, Middle Tennessee State University, Murfreesboro, Tennessee).

Photo 18. George Haynes Slave Cabin – wooden porch on the front with a shed roof, square wooden posts, and a wooden floor (image courtesy the Center for Historic Preservation, Middle Tennessee State University, Murfreesboro, Tennessee).

The reassembled cabin represents the cabin of George Haynes, a man enslaved by Landon C. Haynes [Tipton-Haynes Historic Site Vertical Files]. While the cabin is named after George Haynes, it must be noted that George Haynes never lived in this cabin [Van West, Gavin, Gardner, 2011, p. 63]. This cabin is a substitute for a one and a half story single crib log slave cabin that was originally located on site and that was razed in 1968 due to advanced deterioration. However, the current cabin is located not exactly in the original site of the Haynes` family slave cabin, but only near that place [Tipton-Haynes Historic Site Vertical Files], [Van West, Gavin, Gardner, 2011, p. 64].

Photo 19. George Haynes Slave Cabin at the Tipton-Haynes Historic Site – exterior chimney (image courtesy the Center for Historic Preservation, Middle Tennessee State University, Murfreesboro, Tennessee).

Photo 20. George Haynes Slave Cabin at the Tipton-Haynes Historic Site – interior (image courtesy the Center for Historic Preservation, Middle Tennessee State University, Murfreesboro, Tennessee).

Photo 21. George Haynes Slave Cabin at the Tipton-Haynes Historic Site – fireplace (image courtesy the Center for Historic Preservation, Middle Tennessee State University, Murfreesboro, Tennessee).

Photo 22. George Haynes Slave Cabin at the Tipton-Haynes Historic Site – ladder leading to loft (image courtesy the Center for Historic Preservation, Middle Tennessee State University, Murfreesboro, Tennessee).

According to Van West and Gavin, the chimney and the porch reconstructed on the cabin are incorrect (Photo 18, 19). Most likely a slave cabin would not have had a stone and brick chimney, but instead they would have had a catted chimney. That type of chimney could be pushed away from the house if it caught on fire. Also, in general, slave cabins would not have had a covered porch as was executed at this cabin [Van West, Gavin, Gardner, 2011, p. 147].

Photo 23. Front elevation of the Fox cabin before it was moved to Tipton-Haynes to become the George Haynes Slave Cabin (The George Haynes Cabin Project Records Collection, Tipton-Haynes Historic Site; image courtesy the Center for Historic Preservation, Middle Tennessee State University, Murfreesboro, Tennessee).

Photo 24. Rear elevation of the Fox cabin prior to its translocation to Tipton-Haynes (The George Haynes Cabin Project Records Collection, Tipton-Haynes Historic Site; image courtesy the Center for Historic Preservation, Middle Tennessee State University, Murfreesboro, Tennessee).

Double crib log barn

The double-crib log barn is located northeast of the main house. It was constructed in the early 19th century and it is original to the farm. It has a stone pier foundation, the roof is covered by wooden shingles and there is no chinking between the logs (Photo 25, 27).

The crib on the west side of the barn is one large room with a rough wood floor. It has a wooden ladder allowing access to the interior (Photo 26). The space is currently used for general storage. The crib on the east side of the barn has been divided into two cribs. One of them has a rough wood plank floor and the access to the interior is allowed by a wooden door and a wooden ladder. This part of structure is currently used to store farm implements (Photo 28). The second one has an earthen floor, and is used to store hay [Van West, Gavin, Gardner, 2011, p. 69].

Photo 25.   The double-crib log barn at the Tipton-Haynes Historic Site (photo T. Tomaszek, 2018).

Photo 26.   The double-crib log barn – wooden ladder allowing access to the interior of the crib on the west side of the barn (image courtesy the Center for Historic Preservation, Middle Tennessee State University, Murfreesboro, Tennessee).

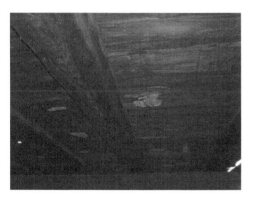

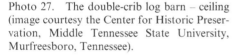

Photo 27.   The double-crib log barn – ceiling (image courtesy the Center for Historic Preservation, Middle Tennessee State University, Murfreesboro, Tennessee).

Photo 28.   The double-crib log barn – interior (image courtesy the Center for Historic Preservation, Middle Tennessee State University, Murfreesboro, Tennessee).

In the 1960s, when the State obtained full possession of the property, the barn was already in very poor condition (Photo 29). In 1965 it was restored under supervision of George Grossman [Baratte, 1970, p. 127]. The structure was documented, photographed and then dismantled. Next, all the reusable logs were numbered. Some of the logs needed to be replaced, and these were made from virgin logs from Unicoi County, Tennessee. After, the barn was rebuilt.

In the next step, all the logs in the structure were numbered again and photographed. Then the barn was dismantled again, and the logs were soaked in a preservative, then the barn was re-built for the final time. In the last stage of the restoration works, a shingle roof was added [Baratte, 1970, p. 127], [Van West, Gavin, Gardner, 2011, p. 70]. As a part of a conservation decision, the exterior boarding from before the restoration was not installed, so the wall of the structure were left uncovered.

Photo 29.   The double-crib log barn before restoration in 1965 (The Gertrude B. Deakins Collection, Tipton-Haynes Historic Site; image courtesy the Center for Historic Preservation, Middle Tennessee State University, Murfreesboro, Tennessee).

The barn later underwent further restoration works. In 1999, the wood shingle roof was replaced, and in 2008 all the drooping log collar beams were replaced [Tipton-Haynes Historic Site Vertical Files], [Van West, Gavin, Gardner, 2011, p. 71].

Corn crib

The log corn crib is located southeast of the barn. It was built in the early 19th century and it is original to the farm. The structure has a stone pier foundation and is covered by a wood shingle roof (Photo 30, 32). The logs were fabricated in the traditional way, so the typical machining marks of hand-processing are noticeable on the surface (Photo 31). As in the case of the double-crib log barn, there is no chinking between the logs. It is a traditional technological solution for this type of farm outbuilding from the early 19th century [Van West, Gavin, Gardner, 2011, p. 72].

Photo 30.   The log corn crib at the Tipton-Haynes Historic Site (photo T. Tomaszek, 2018).

At the upper sections of the cribs there are early 20th century openings, most likely for hay storage. The interiors of the pens have rough wood plank floors. These spaces are currently used for the storage of farm implement artifacts and wood (Photo 33) [Van West, Gavin, Gardner, 2011, p. 72].

Photo 31. The log corn crib – the typical machining marks of hand-processing of the wood noticeable on the surface of logs (image courtesy the Center for Historic Preservation, Middle Tennessee State University, Murfreesboro, Tennessee).

Photo 32. The log corn crib – a stone pier foundation (image courtesy the Center for Historic Preservation, Middle Tennessee State University, Murfreesboro, Tennessee).

Photo 33. The log corn crib – interior of one of the pens (image courtesy the Center for Historic Preservation, Middle Tennessee State University, Murfreesboro, Tennessee).

Photo 34. The log corn crib – deterioration to floor boards in the breezeway (image courtesy the Center for Historic Preservation, Middle Tennessee State University, Murfreesboro, Tennessee).

In the 1960s, when the State gained full possession of the property, the corn crib was in fair condition, so in 1967 the structure was restored by Ralph Nelson and Jerry Bowman (Photo 35). Later it underwent further restoration works, as, among others, in 1999 the wood shingle roof was replaced [Tipton-Haynes Historic Site Vertical Files], [Van West, Gavin, Gardner, 2011, p. 73].

Photo 35.   The log corn crib during period of 1960s restoration of the property (The Mary Hardin McCown Collection, Tipton-Haynes Historic Site; image courtesy the Center for Historic Preservation, Middle Tennessee State University, Murfreesboro, Tennessee).

Pigsty

The pigsty is located northeast of the main house, and to the east of the double crib log barn. This log structure has a stone pier foundation and it is covered by a wood shingle roof. As is typical for this kind of building, there is no chinking/daubing between the logs (Photo 36). The interior has rough wood plank floors (Photo 37). The structure is currently empty [Van West, Gavin, Gardner, 2011, p. 75].

Photo 36.   The pigsty at the Tipton-Haynes Historic Site (photo T. Tomaszek, 2018).

The pigsty is located on the property secondarily as part of the interpretation of the site. It was transferred to the site circa 1971 from another location, possibly in North Carolina [Goist, 1994]. It may dated to circa 1875, however, referring to a report from 1988 by William W. Howell, an architect in Nashville, it also could be erected circa 1871 [Howell, 1988]. According to Russell, this structure is most likely a Smoky Mountains pen type built for protecting pigs from bears, rather than the type of pigsty more commonly associated with the area where property is located [Goist, 1994].

The pigsty was restored in 1997, when a new roof for the structure was raised, and in 1999, when five deteriorated logs were replaced with new ones (Photo 38) [Van West, Gavin, Gardner, 2011, p. 75].

Photo 37. The pigsty at the Tipton-Haynes Historic Site – interior (image courtesy the Center for Historic Preservation, Middle Tennessee State University, Murfreesboro, Tennessee).

Photo 38. The pigsty at the Tipton-Haynes Historic Site – traces of traditional hand-processing of the wood noticeable on the surface of exchanged logs (image courtesy the Center for Historic Preservation, Middle Tennessee State University, Murfreesboro, Tennessee).

Still house

The still house is located east of the main house. This log structure was built in circa 1965 according the traditional building manner and it is a reconstruction of a building that once existed on the property (Photo 39). It has a stone pier foundation and the roof is covered by wooden shingles (Photo 41, 42). The spaced between logs has been chinked and daubed. The original daubing was made of a mixture of mortar, red mud, and dyes, for easier upkeep and to mimic the effect of mud (Photo 40) [Baratte, 1970, p. 128].

The building underwent some later preservation works – in 1985 the exterior door was replaced and in 1999 the roof was replaced. In the mid-1980s, a replica of an 18th century still was placed inside in the still house (Photo 43) [Van West, Gavin, Gardner, 2011, p. 77].

Photo 39.   The still house at the Tipton-Haynes Historic Site (photo T. Tomaszek, 2018).

Photo 40. The still house – traces of traditional hand-processing of the wood used during reconstruction, noticeable on the surface of the logs; daubing made of a mixture of mortar, red mud and dyes to mimic the effect of mud (image courtesy the Center for Historic Preservation, Middle Tennessee State University, Murfreesboro, Tennessee).

Photo 41. The still house – typical covering of roof made of wooden shingles in accordance with the traditional technologies (image courtesy the Center for Historic Preservation, Middle Tennessee State University, Murfreesboro, Tennessee).

Photo 42. The still house – typical covering of roof made of wooden shingles in accordance with the traditional technologies (image courtesy the Center for Historic Preservation, Middle Tennessee State University, Murfreesboro, Tennessee).

Photo 43. The still house – interior (image courtesy the Center for Historic Preservation, Middle Tennessee State University, Murfreesboro, Tennessee).

Spring house

The spring house is located southeast of the still house. This log structure is not original to the site and it dates to circa 1965, when it was built at the time of the first Tipton-Haynes restoration as a reproduction/reconstruction of building existing once at the property (Photo 44).

It has a stacked stone foundation and it is covered by a wooden shingle roof. Like the still house, the spring house was chinked and daubed, most likely following traditional technological solutions. However, the chinking was later repaired in 1985, using non-traditional solution of a mortar coating of six parts brown creek sand, 4 parts hydrated lime, and 1 part mortar mix. The other restoration took place in 1999, during which the roof was replaced [Van West, Gavin, Gardner, 2011, p. 79].

Photo 44.   The spring house at the Tipton-Haynes Historic Site (photo T. Tomaszek, 2018).

The interior of the spring house has no floor and it is opening directly to the spring beneath (Photo 45).

Photo 45.   The spring house at the Tipton-Haynes Historic Site – interior (image courtesy the Center for Historic Preservation, Middle Tennessee State University, Murfreesboro, Tennessee).

The spring house obscures the spring wall, which was restored in 1953 by The State of Tennessee and Tennessee Historical Commission (Photo 46) [Van West, Gavin, Gardner, 2011, p. 79].

Photo 46. Restored spring wall, photo from 1953 (The Mary Hardin McCown Collection, Tipton-Haynes Historic Site; image courtesy the Center for Historic Preservation, Middle Tennessee State University, Murfreesboro, Tennessee).

5   THE HISTORIC SAM DAVIS HOME AND PLANTATION

The home of Sam Davis is in Smyrna, Rutherford County, Tennessee. It represents a typical middle-class house on an average sized farm of the antebellum period, therefore, as an architectural complex in a historical setting together with all its dependencies and land, it is very representative of this kind of properties from that time (Figure 3) [National Register of Historic ..., 1969, p. 2]. It is also significant because Sam Davis lived there before he went away to military school and then to the war which led to his hanging as a spy in 1863.

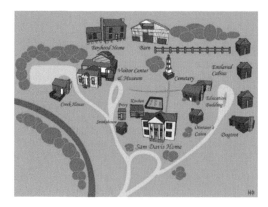

Figure 3.   Schematic map of The Historic Sam Davis Home and Plantation (courtesy the Center for Historic Preservation, Middle Tennessee State University, Murfreesboro, Tennessee).

In fact, two historic homes are located on the farm grounds, one mentioned above and the second one, the log structure which is Sam Davis' boyhood house (Photo 47). The

Photo 47. The Sam Davis` log Boyhood Home at the Sam Davis Home Historic Site (photo T. Tomaszek, 2018).

latter was relocated from the former farm of Davis` family and is now a part of the interpretation of the site[i].

Sam Davis – life and legacy

Called a "Boy Hero of the Confederacy", Sam Davis was born on October 6, 1842. The Davis family initially lived in a log home on their old farm near Smyrna, in Stewartsboro, on Almaville Road near present-day Interstate 24. In 1847, Sam's father, Charles, with a growing family, purchased a larger farm a few miles away. This is the property is now designated as the Sam Davis Home and Plantation Historic Site [Meredith, 1965, p.304].

At age nineteen Sam Davis left home and attended the Military Academy in Nashville to complete his education. In 1861, when the American Civil War began, he joined the Confederate Army in a local militia group mustered into active service as Company I, First Tennessee Infantry. In 1862 he became a member of a company of cavalry known as "Coleman`s Scouts", a group that was meant to work behind enemy lines to disrupt Union communications. The Scouts usually wore Confederate uniforms and traveled with special passes, although if they were captured, they were considered as spies by the Union army and usually then executed [Craddock, 2017].

In the fall of 1863 Davis, with some other Scouts, was assigned to gather information regarding Federal forces movement in Tennessee. On November 20, 1863, while scouting near Pulaski, Davis was captured by Federal soldiers in Giles County while carrying some important documents that were supposed to be delivered to army headquarters (these documents included detailed drawings of Union fortifications at Nashville) [Harcourt, 2006, p. 35]. He was charged with spying and carrying mails to persons in arms against the United States. Even though Davis pleaded not guilty to charge of spying, he was found guilty by martial court on both counts and was hung by Union forces on November 27, 1863 [Craddock, 2017]. After his execution the body of young private was carried back to Smyrna and on December 24, 1863, he was buried in the family burial ground.

---

i. The Historic Sam Davis Home and Plantation: A Nonprofit Organization. Undated. The Sam Davis Memorial Association. https://www.samdavishome.org (accessed March 23, 2018)

Family house and the farm – description

The main building, the so-called Sam Davis` Home, is a two-story structure covered by white clapboard (Photo 48). It is constructed of red cedar logs covered with yellow poplar weather-boards. It is adorned in the portico with four square columns and a balcony, reflecting Greek Revival architecture. The current size and the shape of the house are the result of historical remodeling [Baird, 1956, p. 26]. Apart from being an important example of antebellum architecture, the house in itself has no bigger architectural significance [Meredith, 1965, p.305].

Photo 48.   The Sam Davis` Home at the Sam Davis Home Historic Site (photo T. Tomaszek, 2018).

According to architectural historians, the house originally was constructed as a log structure around 1810 by Moses Ridley, evidence of which partially remained in the front part of the building. It was originally a two story, four-room house with a central hallway. In 1847 the house was bought by Charles Davis, who made changes to it according the architectural fashion of Southern plantation houses at that time (which imitated the Greek Revival style). Later, Charles Davis also covered the logs with poplar siding and enlarged the house by adding the rear wing with its family room, dining room, and sleeping quarters [National Register of Historic . . . , 1969, p. 2].

The building underwent a few more architectural changes in later periods and at present it contains eight large rooms with fireplaces and two stairways that lead to separate sections of the second floor (one side for the boys, the other for the girls). Additionally, a large ell shaped porch joins the original portion of the house with the later north wing. Apart from the sunny rooms in the southwest corner of the house, all of them have at least two windows and, what is characteristic, almost all rooms located downstairs can be entered from the two directions. Many of the original furnishings and belongings from Sam Davis' time are preserved in the house [Meredith, 1965, p.308].

The house is located on a 168 acre antebellum middle-class farm. It is particularly historically important as it represents a typical farm of the period. According to the National Register Nomination Form for the Sam Davis Home, even though many other homes from that time exist and many of them are of finer architectural features, few of the middle-class group of buildings remain in fact in as good a condition as does this farm as a whole. Thus, apart from being the home of Sam Davis, that is why the property is so significant for the history of Middle Tennessee [National Register of Historic . . ., 1969, p. 3].

There are several outbuildings located on the farm grounds, most of which are log structures, as well as log slave quarters (Photo 49). Some of the structures are original to the place, like the kitchen and smokehouse to the left rear of the main house (Photo 50), located nearby to them is the one-room overseer`s office (Photo 51) and the frame barn (Photo 52). Contrastingly, the slave cabins have been rebuilt on the

Photo 49.   The log slave quarters on the grounds of the Sam Davis Home Historic Site (photo T. Tomaszek, 2018).

Photo 50.   The kitchen and smokehouse at the Sam Davis Home Historic Site (photo T. Tomaszek, 2018).

Photo 51.   The one-room overseer`s office at the Sam Davis Home Historic Site (photo T. Tomaszek, 2018).

Photo 52.   The frame barn at the Sam Davis Home Historic Site (photo T. Tomaszek, 2018).

sites of the original quarters. Now some of them are even refurnished [National Register of Historic ..., 1969, p. 3]. The family burial ground, where Sam Davis is buried, is also on the property.

In 1927, the State of Tennessee bought the home from Davis descendants to preserve it as a shrine to the "Boy Hero". Three years later, in 1930, the Sam Davis Memorial Association was created in order to restore and maintain the house [National Register of Historic ..., 1969, p. 3].

Sam Davis log boyhood home at the Sam Davis Home Historic Site

After the Sam Davis Home and Plantation was established as a Historic Site, as a part of the interpretation of the place where Sam Davis lived and grew up, and due to the construction of a subdivision road being built where it stood originally, the former house of Davis` family (Photo 53) was relocated from the original property and placed at the historic site (at the back side of the current Sam Davis Home museum). Therefore, now there are two houses

Photo 53. The log Boyhood Home of Sam Davis at the Sam Davis Home Historic Site (photo T. Tomaszek, 2018).

located on the property that can be linked to Sam – the oldest one, transferred to the site – where the Confederate hero was born and lived in as a boy before his parents moved to the plantation house in Smyrna, where young Sam resided before joining army. The first house in which the Davis family lived is a two-story log cabin, and it was originally built in Ruther-ford County, Tennessee (the same county where The Historic Sam Davis Home and Planta-tion is located), near the intersection of Almaville Road and Interstate 24 in Stewartsboro, just a few miles away from its current location [Meredith, 1965, p.304].

Today a new house sits on the Stewartsboro property. It is occupied and owned by Sherry Maraschiello. The cluster of homes surrounding Maraschiello's property is known as Sam Davis Meadows [Degennaro, 2018].

The Sam Davis Boyhood Home (the relocated first house of Davis family), is basically a combination of two buildings connected by a common chimney – a two-storey residential wing and a separate kitchen. After relocation the structure was never fully preserved and slowly went into disrepair. Its current condition can be described as very bad, and can be observed principally from the front side, where the roofed two-story porch (Photo 54) has been completely destroyed (also partially due to a tornado) (Photo 55).

Photo 54. The two-story roofed porch of the log Boyhood Home of Sam Davis before it was destroyed during the tornado (image courtesy the Center for Historic Preservation, Middle Tennes-see State University, Murfreesboro, Tennessee).

Photo 55. Remains of the two-story roofed porch of the log Boyhood Home of Sam Davis (photo T. Tomaszek, 2018).

To this day, the Sam Davis Boyhood Home has preserved several original, traditional technological solutions, although some of the subsequent modifications have introduced secondary technologies that are detrimental to its original character. Both parts of the building are embedded on regularly spaced masonry pillars made of various sizes of flat stones (rubblestones), combined with a mortar of most probably a secondary composition (unfortunately it was not possible to determine whether and what kind of mortar was originally used to connect the stones of the pillars) (Photo 56, 57, 58)

The embedding of the building on regularly spaced pillars, on which, then, the two "sill logs" were directly laid, is a typical and traditional type of foundation used historically for log structures in this part of the US. The primary factors affecting this kind of foundation construction are climate and the intended permanence of the structure. When the house was erected as a more permanent log dwelling than simple and temporary log cabins, and in order to prevent accelerated wood decay in a warm, humid climate such as in the South, the use of stone piers was common, as it allowed air to circulate beneath the sill logs and keep the foundation drier [Bomberger, 1991].

Examinations determined that the wooden structural elements of the Sam Davis Boyhood Home are mostly original. This is evidenced by the traditional surface treatment of individual

Photo 56. Embedding of the two-storey residential part of the log Boyhood Home of Sam Davis on masonry pillars made of various sizes of flat stones combined with a mortar (photo T. Tomaszek, 2018).

Photo 57. Embedding of the kitchen of the log Boyhood Home of Sam Davis on masonry pillars made of various sizes of flat stones combined with a mortar (photo T. Tomaszek, 2018).

Photo 58. Embedding of the log Boyhood Home of Sam Davis on masonry pillars made of various sizes of flat stones combined with a mortar – detail (photo T. Tomaszek, 2018).

logs. Nevertheless, additions resulting from subsequent repairs of the building are also noticeable. Some of them are made correctly and in accordance with traditional methods of exchanging degraded elements for new ones (Photo 59, 60), although there are also partial fillings using Portland cement in places with degraded wood tissue (Photo 61). This way of "repairing" wooden elements is not only technologically secondary, but above all incorrect, because Portland cement tends to shrink and develop so-called hairline cracks, as well as retain moisture, all of which can potentially damage the logs [Bomberger, 1991].

The technology of filling the spaces between individual logs, which was applied during the re-assembly of the structure after it was relocated to the new site, repeated the traditional solutions[ii]. It is worth mentioning that due to the small width of the voids between

Photo 59. The log Boyhood Home of Sam Davis – fragments with elements which replaced the degraded parts of logs (photo T. Tomaszek, 2018).

Photo 60. The log Boyhood Home of Sam Davis – fragments with elements which replaced the degraded parts of logs (photo T. Tomaszek, 2018).

Photo 61. The log Boyhood Home of Sam Davis – fragments with fillings done with Portland cement in the place of degraded logs (photo T. Tomaszek, 2018).

---

ii. According to Bomberger, for chinking and daubing, there were a variety of different materials, including whatever was most conveniently available. Generally, though, it is usually a three-part system, which was applied in several steps. Most typical chinking consists of two parts. Firstly,, a dry, bulky, rigid blocking, such as wood slabs or stones was inserted into the spaces between the logs. Then, secondly, it was followed by a soft packing filler such as oakum, moss, clay, or dried animal dung. Finally it was covered by daubing, which is the outer wet-troweled finish layer of varying composition, however typically it consist of a mixture of clay and lime or other locally available materials. [Bomberger, 1991].

some of the logs as a chinking, builders usually used stones or wooden tiles placed diagonally (Photo 62). The condition of the chinking and daubing layer is very bad[iii].

The building has two external chimneys on the shorter side elevations. One of them, made of stone and repeating original solutions (reconstructed after translocation entirely from the original material) is external to the entire architectural body (Photo 63), while the other, made incorrectly from brick and incompatible with the original technology, is the connector between the larger element of the residential wing and the smaller kitchen space (Photo 64). The method of reconstruction of the second chimney significantly affected the loss of the original aesthetics of the building and the falsification of traditional building solutions.

Photo 62.   The log Boyhood Home of Sam Davis – the flat lying stones were introduced as chinking (photo T. Tomaszek, 2018).

Photo 63.   The log Boyhood Home of Sam Davis – external chimney made of stone (photo T. Tomaszek, 2018).

Photo 64.   The log Boyhood Home of Sam Davis – external chimney reconstructed in brick (photo T. Tomaszek, 2018).

---

iii. Traditionally, patching and replacing daubing on a routine basis was a seasonal intervention. This was due to environmental factors, like seasonal expansion and contraction of wood, and natural moisture infiltration followed than by freeze-thaw action causing cracks and loosens daubing. [Bomberger, 1991].

Log slave cabins

Once there were numerous slave quarters on the plantation. Like most Southern slave dwellings, they were of quite small size, most likely not bigger than 100 square feet (Photo 69). Some of the cabins had adjacent gardens in which the slaves could grew crops and vegetables. In the late 1920s the original slave cabins from the Davis farm were sold to raise funds to keep the site open. Today the property contains only four historical slave cabins (Photo 65). These, however, were acquired and relocated from other plantations in the process of later interpretation of this historic site by the Sam Davis Association. And so, three of the log structures were moved to their present location from "Rattle and Snap", a large Tennessee plantation near Columbia; and a fourth was moved from Lascassas in Rutherford County, Tennessee [Curtis, 1996, p. 232].

Three of the cabins are one-room structures (Photo 66), and one is a dogtrot, with two rooms and an open central area (Photo 67, 68). The structures have wooden floors and are furnished to present the different types of work done during the antebellum era. One of them is set up as it would have been in the 1850s-1860s, and the two in the middle contain weaving and spinning items and old farming machinery. Some records show that at the peak period in the late 1850s, the Davis Plantation had about fifty-one enslaved who lived and labored on this site, mostly working on cotton fields [Curtis, 1996, p. 232]. According

Photo 65.   Four historical slave cabins on the grounds of the Sam Davis Home Historic Site (photo T. Tomaszek, 2018).

Photo 66.   One of the three one room cabins on the grounds of the Sam Davis Home Historic Site. (photo T. Tomaszek, 2018).

Photo 67.   A dogtrot style slave cabin with two rooms and an open central area located on the grounds of the Sam Davis Home Historic Site – front (photo T. Tomaszek, 2018).

Photo 68. A dogtrot style slave cabin with two rooms and an open central area located on the grounds of the Sam Davis Home Historic Site – back (photo T. Tomaszek, 2018).

Photo 69. One of the log slave cabins on the grounds of the Sam Davis Home Historic Site – a small size of the dwelling compared to the human body (photo T. Tomaszek, 2018).

the other sources, by 1860 the farm was home to 52 enslaved, 27 males and 25 females, who were living in the 14 slave cabins located on the property [Meredith, 1965, p. 315].

Relocated cabins are set up similarly to the Sam Davis Boyhood Home, on pillars consisting of various sizes of flat stones (rubblestones), combined with a mortar of most probably a secondary composition. Everything indicates that the stones are original (were used originally as foundation before relocation), which may be confirmed by the traces of manual processing noticeable on their surface (Photo 70). However, the sill logs are laid directly, not on the stones, but on a piece of metal sheet resting on the top of these pillars (Photo 71). This is to protect the sill logs from moisture pulled up capillarily through the mortar used to connect the stones forming the pillars, but it is a secondary technology and falsifies the original solutions.

Photo 70. The stone pillars used for embedding the slave cabins – detail; noticeable the traces of manual processing on the surface of the stones (photo T. Tomaszek, 2018).

Photo 71. The stone pillars used for embedding the slave cabins – detail: a piece of metal sheet on the top of the pillar and below the sill log, used secondarily for protection against moisture pulled up capillarily (photo T. Tomaszek, 2018).

The cabins have wood shingle covered roofs (Photo 72), as well as the external chimneys. The shape and technology of the execution of these chimneys, according to some researchers and some historical sources, is recognized as typical for a 19th-century slave cabin [James,1981], [Jordan, 1985] (Photo 73, 74, 75). Nevertheless, referring to Van West and Gavin, similar cabins had a catted chimneys, a type of chimney that could be pushed away from the house if it caught on fire [Van West, Gavin, Gardner, 2011, p. 147].

Photo 72.   The slave cabin on the grounds of the Sam Davis Home Historic Site – detail: shingle covered roof (photo T. Tomaszek, 2018).

Photo 73.   One of the slave cabins on the grounds of the Sam Davis Home Historic Site – detail: the external chimney (photo T. Tomaszek, 2018).

Photo 74. The slave cabins on the grounds of the Sam Davis Home Historic Site – view on the side with external chimneys (photo T. Tomaszek, 2018).

Photo 75. A dogtrot style slave cabin on the grounds of the Sam Davis Home Historic Site – the side with external chimney (photo T. Tomaszek, 2018).

Examinations allow us to assume that the wooden elements of the cabins are mostly original (Photo 76). On the surface of individual logs, the traces of traditional, manual method of their fabrication are noticeable. During subsequently conducted repair and conservation works, the completely degraded parts of individual logs were filled with Portland Cement (Photo 77, 78, 79). As it has been mentioned already, this procedure is not only inconsistent with the original material and technological solutions, but above all is a threat to accelerated degradation of wood tissue in the places where the insert from Portland Cement was made (due to the accumulation of excess humidity and blockage of natural wood movements resulting from a change in humidity) [Bomberger, 1991]. The destructive effect of this treatment is observable in many places where there is cracking and separation

Photo 76. One of the of the slave cabins on the grounds of the Sam Davis Home Historic Site – detail: the original wooden logs (photo T. Tomaszek, 2018).

Photo 77. One of the of the slave cabins on the grounds of the Sam Davis Home Historic Site – detail: the completely degraded part of log ending filled with Portland Cement (photo T. Tomaszek, 2018).

Photo 78. One of the of the slave cabins on the grounds of the Sam Davis Home Historic Site – detail: the completely degraded part of log filled with Portland Cement (photo T. Tomaszek, 2018).

of the cement insert from the tissue of wood, which results in the cement fillings falling off and further accelerated degradation of the wood (Photo 80, 81, 82).

The filling of the spaces between individual logs (daubing), made after the relocation of buildings and during subsequent conservations, is inconsistent with traditional technologies. That it is a secondary filling is confirmed by the fact of using concealed reinforcement in the form of a metal wire mesh[iv].(Photo 83, 84, 85) Additionally, doubts are also caused by the coloristic and material composition of the daubing, for which coarse-grained sand with small pebbles and Portland Cement were used (Photo 86).

---

iv. Although contrary to the principle of maintaining the authenticity of technology, concealed reinforcement is sometimes allowed by some in the preservation of historical log structures. Referring to Bomberger, concealed reinforcement if used often „include galvanized nails partially inserted only on the upper side of the log to allow for the daubing to move with the upper log and keep the top joint sealed, or galvanized wire mesh secured with galvanized nails" [Bomberger, 1991].

Photo 79.   One of the of the slave cabins on the grounds of the Sam Davis Home Historic Site – detail: the completely degraded parts of logs filled with Portland Cement (photo T. Tomaszek, 2018).

Photo 80.   One of the of the slave cabins on the grounds of the Sam Davis Home Historic Site – detail: cracking and separation of the Portland Cement insert from the tissue of wood (photo T. Tomaszek, 2018).

Photo 81.   One of the of the slave cabins on the grounds of the Sam Davis Home Historic Site – detail: cracking and separation of the Portland Cement insert from the tissue of wood (photo T. Tomaszek, 2018).

Photo 82.   One of the of the slave cabins on the grounds of the Sam Davis Home Historic Site – detail: cracking and separation of the Portland Cement insert from the tissue of wood and noticeable further accelerated degradation of the wood (photo T. Tomaszek, 2018).

According to Bomberger, Portland cement was used in many late 19th and early 20th century log buildings as a part of the original daubing[v] [Bomberger, 1991]. It is unlikely, however, that it would be used originally for slave cabins around the middle of the 19th century.

The use of Portland cement for the restoration of filling between the logs was a typical way of maintaining historical log structures in the second half of the 20th century. In the literature from these years there are mentioned many recipes for the composition of the mass, which was commonly used for restoration of original daubing. As examples, it can be

v. According to Bomberger „Although a small amount of Portland cement may be added to a lime, clay and sand mix for workability, there should not be more than 1-part Portland cement to 2-parts of lime in daubing mixes intended for most historic log buildings" [Bomberger, 1991].

Photo 83. The slave cabins on the grounds of the Sam Davis Home Historic Site – detail: concealed reinforcement in the form of a metal wire mesh used for strengthening the secondary daubing (photo T. Tomaszek, 2018).

Photo 84. The slave cabins on the grounds of the Sam Davis Home Historic Site – detail: concealed reinforcement in the form of a metal wire mesh used for strengthening the secondary daubing (photo T. Tomaszek, 2018).

Photo 85. The slave cabins on the grounds of the Sam Davis Home Historic Site – detail: concealed reinforcement in the form of a metal wire mesh used for strengthening the secondary daubing (photo T. Tomaszek, 2018).

Photo 86. The slave cabins on the grounds of the Sam Davis Home Historic Site – detail: secondary daubing made of coarse-grained sand with small pebbles and Portland Cement (photo T. Tomaszek, 2018).

given the composition of the mass suggested by Donald A. Hutsler[vi] and the composition of the mass included in the book by Harrison Goodall and Renee Friedman[vii].

vi. Hutsler D. A. 1974. *Log Cabin Restoration: Guidelines for the Historical Society*. American Association for State and Local History, Technical Leaflet No. 74, *History News*, Vol. 29, No. 5 (May 1974.)

vii. Goodall H., Friedman R. 1980. *Log Structures: Preservation and Problem-Solving*. Nashville, TN: American Association for State and Local History.

According to the first proposal, the Daubing Mix should look like this [Hutsler, 1974]:

1/4	Cement
1	Lime
4	Sand
1/8	dry color
	hog bristles or excelsior

While, according to the second proposal, the ingredients should be prepared in the following proportions [Goodall, Friedman, 1980]:

1	portland cement
4-8	Lime
7-10	Sand

The other outbuildings

Near the slave cabins there is a log building currently used as an "education center". This structure has an external chimney, partially reconstructed of bricks (Photo 87). The shape and technological method of this reconstruction are inappropriate and falsify original solutions.

Photo 87.   A log building currently used as an "education center" on the grounds of the Sam Davis Home Historic Site; noticeable an external chimney, partially reconstructed of bricks (photo T. Tomaszek, 2018).

The logs of the one-room overseer's office (Photo 88) are original and the traces of hand-manual fabricating of these elements are visible on their surface. However, the daubing, which was restored during subsequent conservations, is not of original recipe (Photo 89). Additionally, as in the case of the log slave cabins mentioned above, the completely degraded parts of individual logs were filled with Portland Cement (Photo 90). This improper conservational solution was also executed on the wooden elements of the smoke-house (Photo 91).

Photo 88. The log one-room overseer`s office on the grounds of the Sam Davis Home Historic Site (photo T. Tomaszek, 2018).

Photo 89. The log one-room overseer`s office on the grounds of the Sam Davis Home Historic Site – detail: the traces of hand-manual fabricating noticeable on the surface of logs and the secondary daubing (photo T. Tomaszek, 2018).

Photo 90. The log one-room overseer`s office on the grounds of the Sam Davis Home Historic Site – detail: fillings with Portland Cement in the places of completely degraded wood (photo T. Tomaszek, 2018).

Photo 91. The smokehouse on the grounds of the Sam Davis Home Historic Site – detail: fillings with Portland Cement in the places of completely degraded wood on the endings of the individual logs (photo T. Tomaszek, 2018).

6 THE ISSUE OF THE RECONSTRUCTION AND TRANSLOCATION OF WOODEN BUILDINGS AS A PART OF THE INTERPRETATION AND PRESERVATION OF A GROUP OF HISTORICAL WOODEN STRUCTURES IN THE EFFORT TO MAINTAIN THE AUTHENTICITY OF HERITAGE SITES

Complexes of heritage wooden buildings constituting a historical site which is interpreting a given historical period, show the essence of the issue of authenticity as the horizon responsible for the genuine transmission of architectural, technological and historical narrative. Thus, analysis of solutions and conservation interventions allows verification of the adopted assumptions and content shown.

The Tipton-Haynes Historic Site is an example of the use of various methodologies and approaches to conservation within one historical site, which despite often conflicting solutions constitute the implementation of one intended goal – the interpretation of a historical farm in a selected historical context. A special form of the implementation of this approach can be observed in the main house, being a conglomerate of parts from different periods, each of which is differently displayed and conserved. The house no longer has the character of a log building, however, its structure retains the original fragments of the log cabin to give some interpretation of the architecture of the original cabin. Therefore, that part of the cabin is preserved as the original wall from the Tipton-era log cabin and it is exposed in one of the rooms.

The Tipton-Haynes Historic Site also allows us to observe the dynamics of the concept of interpretation and resulting from this, consequences in terms of the horizon of authenticity. Even properly based on historical references, this historic site is an example of over-interpretation and a new given context, which took place during the transformation of the site into a Living Farm. All of it was organized to demonstrate the live of the pioneers. Going beyond authenticity, understood as maintaining original architectural solutions and showing the original character of the place, the implementation of this aspect of historical function in the contemporary context of the place turned out to be a kind of narrative of an artificially created secondary situation.

The variety of conservation methodologies mentioned above can be seen primarily in the case of the outbuildings located on the Tipton-Haynes farm, which are an interpretation of the original farm and its original spatial and architectural character. Next to the original buildings, there are architectural objects which are secondary reconstructions and buildings transferred from another place. Interestingly, not all former historical outbuildings which did not survived to this day were reconstructed, but only selected ones. The reason for this and no other selection is unknown and allows us to suppose that it is the result of a free interpretation and inconsistent with maintaining of authenticity of this place.

The smokehouse, the still house and the spring house are complete reconstructions. They are built in accordance with traditional technologies (with traditional solutions and materials), they repeat the original form/shape (they are an exact reconstruction of buildings that once existed on property) and now they are kept with a natural patina. As such, they can be treated as buildings that maintain the authentic character and narrative of the place, assuming that authenticity is understood as the original shape and technology, not as the original fabric.

However, some of the technological solutions used are contradictory to the original ones. And so, the daubing applied on the still house is not of the traditional recipe. It is only an imitation, which is designed to mimic the original effect of mud. A similar situation is with daubing on the spring house. These secondary technological solutions were executed during later conservation works and most likely the daubing used during the time of reconstruction was of original recipe. As such, the secondary technologies have a negative impact on the authenticity of the buildings.

The buildings originally belonging to other places and then relocated to The Tipton-Haynes Historic Site are George Haynes Slave Cabin and the pigsty. The relocation of a wooden building to a new place always raises a number of questions and doubts related to the loss of authenticity of the place and context. In addition, in the case of George Haynes Slave Cabin, its reassembling and the addition of the new parts led to an over-interpretation of its original character, shape and narrative. The building is located not exactly on the site of original cabin, it also gained a new chimney and porch. Both of these elements are not original to the structure, the chimney as a fabric and type (it is historical chimney, but from different building), the porch as architectural element (the cabin originally did not have a porch). Their reconstruction is kind of free interpretation and is done not in accordance with the solutions used during the time of the erection of buildings.

The pigsty is also not an original type of this structure to the site, as most likely it is a Smoky Mountains pen-type built for protecting pigs from bears, rather than the type of pigsty more commonly associated with the area where the property is located.

A separate issue is the interpretation made on the double-crib log barn. Although it is the original structure to the site, its primary character was changed as the original exterior boarding was removed. Thus, it lost its authenticity of original design as well as original solutions, becoming the example of a building showing over-interpretation made quite freely.

The Historic Sam Davis Home and Plantation shows a similar issue of the application of various conservation methodologies on a set of historical wooden buildings, and the resulting implications regarding the issue of authenticity. Just as in the case of the main house of The Tipton-Haynes Historic Site, the Sam Davis Home was originally a log cabin, which later was the basis for secondary architectural transformations and now in the form of historical remains is exposed in the main body of the building.

As a historic site representing the context of a specific historical object, The Historic Sam Davis Home and Plantation is also a monument interpreting the place of residence of an important person in Tennessee history. Consequently, because this interpretation, it is basically a site created because of the secondary narrative and arbitrarily showing the intended ideological content. This is evidenced by the organization of the architectural and urban layout, within the scope of which has been designated the area of the interpretation of enslavement, existing parallel to the interpretation of the childhood of Sam Davis. As part of this intended interpretation, the first home of the Davis family, now called Sam Davis Boyhood Home, has been relocated into the farm. Although it is an original house in which Sam Davis spent the first years of his life, it was transferred in a new context, both spatial and architectural as well as historical. Thus, the example of this building reveals extremely complicated issues of the relocation of a wooden building, done as a form of interpretation of a historical place in the perspective of maintaining the horizon of authenticity of a historic wooden building.

Although the Sam Davis Boyhood Home has been moved into a new spatial context, it is on the farm of the same family that once lived in it, and in the same cultural landscape area. Thus, its context of belonging to the same family and the same geographic context (the same locality) has not changed. The same is true of its original architectural shape, which has not been modified after relocation (although it is now in ruins). The above aspects, as a message of the original form and the original context of belonging, are, however, only elements of a broader interpretation and as such cannot be regarded as authentic. After the relocation, Sam Davis Boyhood Home became an element of a new narrative, for which needs it is located in the present place and fulfills this, and not other functions.

Another issue is the falsification of the original technological solutions of the building, only to recall the rather free reconstruction of the chimney made in brick. Such interventions, which are an optional reconstruction of the visual aspect of the historic building, are inconsistent with the horizon of authenticity and also largely misrepresent the original aesthetics.

The interpretation of slavery, as a result of which four slave cabins originally located in two different places were relocated to the site, established only a general image of the farm from the mid-nineteenth century. The set of architectural objects created in this way which are representing the former presence of slaves on the plantation is a fairly free narration, and it only refers to the original number of cabins, their organization and location, as well as the method of their use. In addition, after the relocation and during subsequent maintenance, secondary technological and aesthetic solutions were used, which seriously affects the loss of the horizon of the authenticity of these objects.

7   CONCLUSIONS

Preserving and maintaining a complex of wooden buildings constituting a historic site is a multidimensional task and often requires the use of various conservation methodologies. It remains problematic whether such procedure and method of protection, in which often contradictory ideological assumptions are implemented, is consistent with the horizon of authenticity of heritage wooden building. The examples discussed showed difficulties resulting from the use of different conservation approaches/procedures and problems related to the reconstruction and translocation of a wooden building into a new spatial and historical context.

The most problematic issue is the use of reconstruction and relocation as a way of interpreting a historic site. The implementation of a new architectural element, even if it is a strict reconstruction of a pre-existing one, is always a challenge to maintaining the horizon of the authenticity of the site. At the same time, the reconstruction process itself faces many challenges, among which should be mentioned, for example, the importance of the repetition of original technological or formal solutions that are necessary from the perspective of the authenticity of the historic building. In the case of relocation, the original spatial context is changed, which significantly affects the modification not only of the authenticity of the location, but also the change of the formal and architectural aspect of the entire site. The same question remains whether the secondary introduced architectural element does not negatively affect the modification of the historical narrative of the place in which it was located after relocation.

ACKNOWLEDGMENTS

The above research was carried out during the years of 2017/2018 under a Research Fellowship at the Center for Historic Preservation, Middle Tennessee State University, Murfreesboro, Tennessee, USA. It was also possible thanks to a research grant from the Kosciuszko Foundation.

Special thanks to Professor Carroll Van West, PhD; Tennessee State Historian, Director of MTSU Center for Historic Preservation and Director of the Tennessee Civil War National Heritage Area, and Stacey Graham, Ph.D.; Research Professor at MTSU Center for Historic Preservation for their help and commitment in carrying out this research.

REFERENCES

American Heritage Trees. Historic Trees from an Historic Farm. 2019. Sam Davis – Historic Sam Davis Home and Plantation. https://americanheritagetrees.org/partners/sam-davis/(accessed February 29, 2019).

Baird M. R. 1956. *The Home of Sam Davis*. W&G Publishing, Nashville, Tennessee.

Baratte J. J. 1970. *The Tipton-Haynes Place: II. The Later Years*. Tennessee Historical Quarterly, XXIX, No. 2 (Summer): 125-129.

Bomberger B. D. 1991. *The Preservation and Repair of Historic Log Buildings*. Preservation Briefs #26, Technical Preservation Services, National Park Service, US Department of the Interior (available at: https://www.nps.gov/tps/how-to-preserve/briefs/26-log-buildings.htm)

Boyd Jr., C. C. 1990. *Archaeological Investigation at the Tipton-Haynes Historical Farm*. Johnson City, Tennessee.

Butcher-Younghans S. 1993. *Historic House Museums: A Practical Handbook For Their Care, Preservation, and Management*. New York: Oxford University Press.

Corlew R. E. 1998. *John Sevier,*/in:/Tennessee Encyclopedia of History and Culture, edited by Van West C., Nashville: Rutledge Hill Press: 838-840.

Craddock P. 2017. Sam Davis. *Tennessee Encyclopedia*, Tennessee Historical Society. https://tennes
seeencyclopedia.net/entries/sam-davis/(accessed February 23, 2018).

Curtis N. C. 1996. *Black Heritage Sites: An African American Odyssey and Finder's Guide*. Library of
Congress, American Library Association, Edward Brothers Inc., Washington D.C., pp. 231-232.

Deed from John R. Banner to Robert W. Haynes, Washington County Deed Book 40, Page 499.

Deed from John White to S.L. Simerly dated August 1, 1872, Washington County Deed Book 46,
Page 316.

Deed from Samuel Henry to John Tipton, dated May 15, 1784. Washington County Deed Book 1,
Page 301.

Deed from Samuel P. Tipton, Elizabeth Tipton and Edna Tipton to David Haines, dated February
28, 1837. Washington County Deed Book 22, Page 177.

Deed from Samuel W. Simerly and Lawson G. Simerly to the State of Tennessee, dated November 25,
1944. Washington County Deed Book 227, Page 344.

Degennaro N. 2018. Property where Civil War boy hero Sam Davis was born is up for sale. Murfreesboro Daily News Journal, Oct 2, 2018. https://eu.dnj.com/story/news/2018/10/01/sam-davis-birth
place-propertysale/1340557002/

Goist D. 1994. *Conservation Assessment Tipton-Haynes Historic Site*. Series X, Box 1, Folder 5, "Conservation Assessment Program grant, 1994", Tipton-Haynes Historical Association Records.

Goodall H., Friedman R. 1980. *Log Structures: Preservation and Problem-Solving*. Nashville, TN:
American Association for State and Local History.

Harber's History Lesson: Sam Davis' brothers also suffered losses. 2015. Murfreesboro Daily News
Journal. May 16, 2015. https://eu.dnj.com/story/news/2015/05/15/harbers-history-lesson-sam-davis-
brother-also-suffered-losses/27414329/(accessed January 17, 2018).

Harcourt E. J. 2006. *"The Boys Will Have to Fight the Battles without Me": The Making of Sam
Davis, "Boy Hero of the Confederacy"*. Southern Cultures, Fall 2006, pp: 29-54.

Howell W. W. 1988. *Tipton-Haynes Home, Volume 15* – Management Summary.

Hutsler D. A. 1974. *Log Cabin Restoration: Guidelines for the Historical Society*. American Association for State and Local History, Technical Leaflet No. 74, "History News", Vol. 29,
No. 5 (May).

James P. 1981. *Architecture in Tennessee. 1768-1897*, Knoxville: The University of Tennessee Press.

Jordan T. G. 1985, *American Log Buildings: An Old World Heritage*. Chapel Hill, NC: The University
of North Carolina Press.

Lawson D. T. 1970. *The Tipton-Haynes Place: I. A Landmark of East Tennessee*. Tennessee Historical
Quarterly, XXIX, no. 2 (Summer), pp. 105-124.

Meredith, O. N. 1965. *The Sam Davis Home./*in:/Tennessee Historical Quarterly, XXIV, pp. 303-320;
Tennessee Historical Society, Nashville, Tennessee.

National Register of Historic Places Inventory – Nomination Form for Sam Davis Home, Smyrna,
Rutherford County, Tennessee; United States Department Of The Interior, National Park Service;
May Dean Coop, Tennessee Historical Commission, Nashville, Tennessee; 31 July 1969.

Specifications for restoration of the Main house. Undated. Series VI, Box 1, Folder 17, *Restoration
Work (1965-1970)*, Tipton-Haynes Historical Association Records.

Stahl R. 1986. *Greater Johnson City: A Pictorial History*. Norfolk/Virginia Beach: Donning Com-
pany/Publishers.

The Historic Sam Davis Home and Plantation: A Nonprofit Organization. Undated. The Sam Davis
Memorial Association. https://www.samdavishome.org (accessed March 23, 2018).

Thomas R. C. 1989. *Are You a Descendant of Landon Carter Haynes?*. Ansearchin News 36, No. 2
(Summer), p. 57.

Tipton Haynes. National Register of Historic Places nomination, prepared by Harper H. L., National
Park Service, Tennessee Historical Commission, January 21, 1970.

Tipton-Haynes „living farm" proves a new concept in historical sites, Series VII-A, Box 1, Folder 10,
Newspaper Clippings (1970), Tipton-Haynes Historical Association Records.

Tipton-Haynes Historic Site Vertical File, Archives of Appalachia, East Tennessee State University,
Tennessee.

*Tipton-Haynes Historic Site: Correspondence,/*in:/Inventory of State Land, McCown Collection, dated
November 24, 1960, Box 7, Folder 8.

Tipton-Haynes Historical Site: Restoration of Buildings (1966-1967), Deakins Collection.

Van West C. 1998. *John Tipton./*in:/Tennessee Encyclopedia of History and Culture, edited by Van
West C., pp. 980-981, Nashville: Rutledge Hill Press.

Van West C., Gavin M., Gardner Leigh A. 2011. *Tipton-Haynes Historic Site – History, Conditions Assessment & Maintenance Recommendations*. Tennessee Historical Commission (THC), the Tipton-Haynes Historic Site, and the Tennessee Civil War National Heritage Area (NHA), MTSU Center for Historic Preservation, Murfreesboro, Tennessee.

Waterfield C. W. Jr. 1967. *Tipton-Haynes Historic Site: Tipton Haynes House (1967)*. AIA, Report dated July 21, 1967, Box 1, Folder 15, Deakins Collection.

Whittle D. 2014. Slave`s story disclosed at Sam Davis Home. *The Cannon Courier*, February 5, 2014. http://www.cannoncourier.com/slaves-story-disclosed-at-sam-davis-home-cms-11892 (accessed March 23, 2018).

Williams S. C. 1970. *History of the Lost State of Franklin*. Johnson City: The Watauga Press, 326.

Wills W. R. 1986. *The Belle Meade Farm: Its Landmarks and Outbuildings*. Nashville Chapter Association for the Preservation of Tennessee Antiquities, Nashville, Tennessee.

Wilson M. 1984. *Log Cabin Studies*. Cultural Resources Report No. 9. Ogden, UT: United States Department of Agriculture, Forest Service.